Journal of the Architects Regional Council Asia (ARCASIA)

U0166580

Copyright © 2021 Tongji University Press

Architecture Asia is published quarterly. Reproduction in whole or part
without written permission from the Publisher is strictly prohibited.

Although every effort has been made to ensure accuracy in the
preparation of each publication, the publisher, printer and editorial team/
staff accept no responsibilities from any effects arising from errors or
omissions. Opinions expressed in the publication are those of the
contributors and not necessarily endorsed by the publisher, printer and
editorial team/staff.

Logo Design: JULY DESIGN GROUP

Book Design: Telos Books

Cover Project: Morning Dew Guesthouse by Architects Group RAUM,
Republic of Korea

National Library of China Cataloguing-in-Publication Data. A catalogue
record for this book is available from the National Library of China ISBN
978-7-5608-8649-7

Printed and Bound in the People's Republic of China.
ISSN 1675-6886

Contents

Editorial

In the process of globalization, the governments of Asian countries have developed various programs to attract foreign investment and stimulate local economic growth through urbanization. As a result, the last thirty years have witnessed rapid urban growth in Asian countries. New urban areas are being built at an astonishing speed and the number of rural migrants to cities is increasing sharply. Consequently, cities face tremendous pressure in accommodating the burgeoning population and economic activities. These trends have reshaped the industrial structure, spatial form, daily urban life, and rural–urban relationships of Asian cities in significant ways.

However, in the past decade, a crisis in globalization has hindered the cross-border movement of capital and dramatically reduced the exchanges of information and people between Asia and other parts of the world. These significant changes to the globalization process may result in a new direction for urban development in Asia. At this crucial juncture, it is important for us to reflect on the decades-long patterns of rapid urban growth in Asia and consider how to transfer the basis of Asian urban development from "construction" to "regeneration."

To respond to this issue, the 3rd ARCASIA International Conference on Urbanism (Shanghai, December 2020) focused on "urban regeneration." The conference sought various perspectives and experiences from different regions around the world through three themes: regeneration and urban heritage, regeneration and the city, and regeneration in a new perspective.

This issue contains articles from five speakers which illustrate global research on urban regeneration. Nilda Valentin proposes the urban regeneration of industrial sites and their valorization and reuse as an alternative to urban sprawl. The study examines the urban area of Rome, Italy, where urban growth is no longer sustainable and the use of abandoned industrial sites is critical. Samra Mohsin Khan describes the urban development and renewal of Islamabad, the capital of Pakistan, since the 1960s. The article provides an introduction to the planning history of Islamabad, analyzes in detail the specific problems it is currently facing, and reflects on the sustainable development of Islamabad while preserving its history and character. Feng Lu, Feng Li and Wu Jiao provide a comprehensive overview of the significant historical changes undergone by "Columbia Circle" in Shanghai over the past century. The three phases of change reflect three concepts of the ideal place of the city. Yao Dong's article introduces the possibility of socializing the renewal of the dense and aging neighborhood through a special form of combined practice–teaching. The article highlights how the service-learning team became involved in community planning and filmed the transformation over a whole year. Dong Nannan and Zhao Shuangrui's article introduces ecological regeneration through vertical greening. It analyzes the latest skyrise greenery projects in the urban renewal of Shanghai from the integrated aspects of planning tools, sustainable design, efficiency evaluation, economic benefits, and smart maintenance technology.

Industrial Heritage in Rome: Valorization and Reuse of Abandoned Industrial Areas

Nilda **VALENTIN**, Sapienza University of Rome, Italy

Abstract

In recent decades, abandoned industrial sites have emerged in areas affected by deindustrialization. The lifecycle of buildings, from construction to operation to abandonment, is simultaneous with substantial changes affecting urban and natural landscapes and quality of life. This paper focuses on the urban regeneration of industrial sites and their valorization and reuse as an alternative to urban sprawl. The author applies her architectural and academic background to describe application studies within the urban area of Rome, Italy, where urban growth is no longer sustainable and the presence of abandoned industrial sites is critical. A methodology is proposed to address this challenge and select optimal architectural and urban renewal strategies. This process involves a multilayered ethical/aesthetic assessment of cultural, environmental, economic, and social issues. In-depth studies of heritage sites are illustrated; these demonstrate how the proposed methodology results in greater resiliency, flexibility, and sustainability, thereby improving the area's quality of life.

Author's Information
Nilda VALENTIN: nilda.valentin@uniroma1.it, ORCiD: 0000-0001-9765-4462

Keywords

Industrial heritage, protection, valorization, reuse, urban regeneration, Rome.

Figure 1
Luciano Cupelloni, MACRO
Future Testaccio Museum of
Contemporary Art, Rome,
2002–07.

1. Introduction

Abandoned industrial sites have appeared in areas affected by deindustrialization and delocalization in recent decades. The lifecycle of buildings, from construction to operation to abandonment, occurs concurrently with other changes that affect the urban and natural landscape and an area's quality of life. The need for an in-depth understanding of the architectural and urban problems and the social, economic, and environmental conditions of former industrial sites has increasingly become the focus of architects, urban planners, and policymakers in several countries who are interested in industrial heritage regeneration.

Various architectural and urban planning strategies have been used to address the complex issues associated with abandoned industrial sites. These strategies range from the total demolition of existing structures and construction of new facilities to the partial or total protection, valorization, and reuse of industrial heritage sites. Urban sprawl and further land consumption are avoided in the latter strategies, while the valorization and reuse of heritage sites can revitalize cultural assets and provide a means of recovering historical memory, identity, and the character of a place, providing significant added value to the urban regeneration project.

The re-functionalization of former industrial buildings with new cultural, social, and economic activities can generate a new sustainable reality in places that are considered promising for future urban and territorial transformation. The reuse of industrial heritage, favoring the regeneration of entire parts of a city, offers the opportunity to redesign the city and improve its socioeconomic condition.

The following analyses of redevelopment projects implemented in Rome illustrate how the image of the city has changed over the past 25 years and highlight the strategies used for the regeneration and reuse of industrial heritage.

2. Industrial Regeneration and Reuse in the City of Rome

In Rome, after a long period of focus on the city's oldest historical assets, public awareness of urban growth and the presence of several abandoned industrial areas led the municipal government to encourage the regeneration of industrial sites. Several studies and analyses have been undertaken to identify critical industrial areas, most of which are now designated in the current general master plan of the city.

Redevelopment plans have been adopted to define and enhance industrial heritage sites, reorganize the surrounding urban fabric, and develop the associated infrastructure. Although a top-down approach has prevailed in the proposals, citizen participation and advocacy have helped to achieve optimal solutions and ensure project sustainability. The participatory process, based on public meetings, open workshops, and online surveys, has enabled the identification of specific community needs and the formulation of agreed-upon design strategies for the regeneration of industrial sites. The rehabilitation and revitalization of entire marginalized and economically stressed urban areas with the support of citizen participation and consent allows the generation of more resilient and sustainable cultural, creative, and innovative activities. This process can transform unresolved sites into areas that provide better quality of life.

In recent decades, several major international architectural competitions for the redevelopment of industrial heritage sites have been held in Rome. Most of these sites have become important cultural nodes, contributing to the decentralization of activity, historic center decongestion, and the creation of several public

3

Figure 2
Centrale Montemartini
Museum, Rome, 2005.

Figure 3
Former gasometer, Rome,
1935–37.

areas where people can meet and gather.

In 2000, for example, an urban redevelopment plan was adopted for the southern Ostiense–Marconi district. The plan included abandoned industrial areas such as the former slaughterhouse (ex Mattatoio), the former thermoelectric power plant (Centrale Montemartini), the former general market (ex Mercati Generali), the former Miralanza factory, and the former gasometer. The primary urban strategy was to create a new education and cultural hub hosting a public university, schools, exhibition facilities, and institutional services.

The ex Mattatoio of Testaccio, built in 1888–91 by Gioacchino Ersoch and decommissioned in 1975, now houses several cultural, educational, and social spaces such as the Academy of Fine Arts, the Faculty of Architecture Roma Tre, the MACRO Future Testaccio Museum of Contemporary Art (Figure 1), a music school, and the Global Village. Historic buildings have been renovated to accommodate the new functions, with original machinery often in view that evokes the area's past.

The large size of the buildings has enabled the flexible use of space, such as the conference room designed for the university, an example of "architecture within architecture." Additions using modern materials and technologies have been made in harmony with the preexisting buildings. The new green areas and squares provide places in which students and visitors can socialize. The large footprint of the complex and its multiple functions have made it "a city within the city."

The former Centrale Montemartini thermoelectric plant, which operated from 1912 to 1963, is a unique museum where classical Roman sculptures are placed next to original machinery, creating an evocative and original dialogue between archeology and industrial heritage. The structure was inaugurated in 1997 as a temporary exhibition space during the renovation of the

Capitoline Museums. The successful mix of culture and technology led to it becoming a permanent museum in 2005 (Figure 2).

The ex Mercati Generali, designed by Rem Koolhaas in 2005, aimed to create a "youth city" (Città dei Giovani); however, after 15 years, the project is still under construction. The discovery of Roman ruins, budget problems, and functional changes have caused continuous work interruptions and Koolhaas' resignation. The proposal aims to preserve the exterior historic buildings while enclosing and obstructing the new interior architecture from view.

Many industrial buildings in the neighborhood have been repurposed as offices, residential, and commercial spaces, and others are awaiting renovation, such as the former gasometer and Miralanza factory. The gasometer, built between 1935–37, is now the symbol of the entire district (Figure 3), while the Miralanza factory, open from 1899 to 1952, has been partly demolished to make room for future developments. The neighborhood committee has proposed the development of an urban park in its place. The old warehouse now houses a community theater, but the remaining buildings continue to deteriorate.

In 2016, a nonprofit cultural organization dedicated to urban creative experimentation sponsored an outdoor art exhibition created by the French artist Seth in the abandoned buildings. The exhibition aimed to bring the attention of public and private stakeholders to industrial heritage. The bottom-up initiative has now established an unofficial museum called Museo Abusivo Gestito dai Rom at the site. A recent architectural competition has led to the redevelopment of the area through private involvement only.

In 2009, a similar creative bottom-up intervention took place at the abandoned Fiorucci factory on Via Prenestina to draw attention to the occupants' economic and housing problems. Here, the Museo

Figure 4
Zaha Hadid, MAXXI, Museum
of 21st Century Arts, Rome,
2010.

Figure 5
Odile Decq, MACRO, the
Museum of Contemporary Art
of Rome, 2010.

dell'Altro e dell'Altrove di Metropoliz was born as an informal street art exhibition space, with works created for free by leading artists from around the world. These creative approaches are now being studied to identify the best way in which to revitalize the aforementioned industrial sites. The Ostiense-Marconi district is now a central cultural hub in the southern part of the city.

In the northern Flaminio district, a cultural approach has also been used to revitalize an area characterized by old military barracks and the 1960 Summer Olympic Village. The primary urban planning strategy has been to create a "cultural axis" that runs from the 2011 Ponte della Musica (Music Bridge), designed by Buro Happold, to the 2002 Auditorium Parco della Musica (Music Park), a complex with three concert halls and an outdoor theater, designed by Renzo Piano.

Other structures are found along the cultural axis, such as the 1957 Palazetto dello Sport designed by Pierluigi Nervi, a community theater, and the MAXXI, the 2010 National Museum of 21st Century Arts designed by Zaha Hadid (Figure 4). The museum consists of new fluid architectural volumes that embrace and intersect the old military barracks, outlining a large public square that connects the two opposing sides of the urban block. Although some of the buildings were demolished, the urban fabric's rupture has encouraged community socialization and new cultural and economic activity in the surrounding area. This has revitalized the entire neighborhood, which has become the city's primary northern hub.

MACRO, the Museum of Contemporary Art of Rome, designed by Odile Decq and opened in 2010, is a cultural center located in the eastern part of the city in the former Peroni brewery on Via Nizza. Here, the old industrial complex's main building has been completely emptied and now houses a new steel and glass structure within the existing external walls. The museum is characterized by its roof garden and the sculptural red interior volume, which contains the main conference room. The glass corner marks the main entrance and the pedestrian walkway crossing the building to connect the urban block's opposite sides (Figure 5). This unique architectural promenade allows the entire complex to be open to the neighborhood.

Many other areas in Rome also need regeneration. Owing to the current economic crisis, many heritage redevelopment projects are now entrusted to the private sector. Nevertheless, public government has the responsibility to approve proper architectural and urban planning solutions and to support citizen participation in the decision-making process.

The redevelopment strategies used at former industrial sites have allowed the city to develop new cultural assets while maintaining the historical memory and identity of places. They have also promoted social activity and created new economic opportunities. For these reasons, it is essential to explore the regeneration and reuse of industrial sites to recycle heritage and revitalize entire neighborhoods.

3. Industrial Heritage in Practice: Proposals and Built Projects

Over the years, industrial heritage has been included in some architectural proposals and realizations for the city. In the study carried out for the new master plan of Rome in 2000, five urban areas characterized by the presence of abandoned industrial buildings were identified, analyzed, and provided with specific solutions for their redevelopment and infrastructure problems (Figure 6).

For the urban redevelopment of Borghetto Flaminio, located along the city's central north–south axis, the idea was to create a cultural and artistic center that could resolve the neglected relationship between the historic center and the

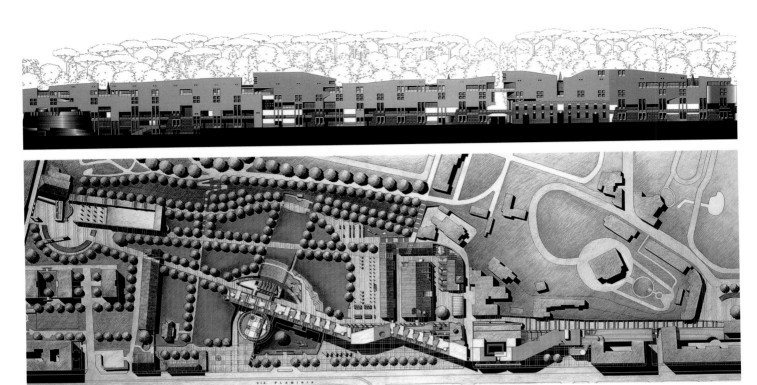

Figure 7
Lenci-Valentin, proposal for
the urban redevelopment of
Borghetto Flaminio, Rome.

system of museums and academies of Valle Giulia. The project area is characterized by the prominent north–south axis of Via Flaminia, the wooded cliff to the east that borders Villa Strohl-Fern, the Villa Vagnuzzi (home of the Accademia Filarmonica Romana), the Roman ruins, and the former public transportation depot. This area, located beyond Porta del Popolo, is one of the rare places where the tuffaceous ridge of Villa Borghese is still visible, an image so frequently represented in the travel journals of the artists who visited the capital from Via Flaminia. The design strategies aim to protect the existing Roman ruins, reuse the industrial buildings, enhance the natural landscape, create an articulated mixed-use building that, breaking the nineteenth-century street alignment, could create new road dilatations, pedestrian networks, and open space for socializing (Figure 7).

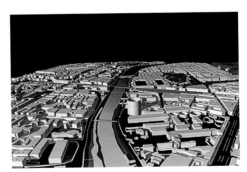

Figure 6
Lenci-Valentin, study for the
new master plan of Rome
in 2000.

The 2000 International Competition for the Transformation and Urban Renewal of the San Lorenzo Area, focused on developing a newly built system above the railway tracks of Termini Station, with new services and pedestrian connections designed to mend the broken urban fabric. Office, educational, commercial, governmental, and recreational facilities are housed in the new high-rise, mid-rise, and bridge buildings, creating a *mixité* capable of bringing new social and economic activities to the fragmented neighborhood. The architectural volumes were designed to transform the visual order in a city sector that was previously disrupted by railroad tracks. On an urban and architectural scale, the entire development was designed to play a key role in defining the place's sustainability, recognizability, and identity while generating new economic opportunities (Figure 8).

The New Cultural Center of Pomezia, currently under construction, consists of a new 450-person theater and an

Figure 8
Lenci-Valentin, proposal for
the transformation and urban
renewal of San Lorenzo,
Rome.

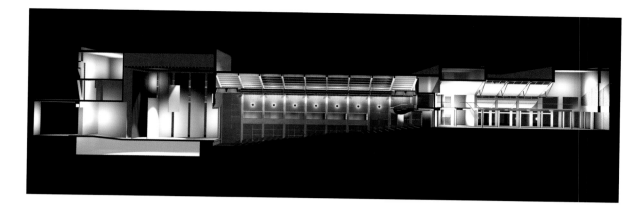

Figure 9
Petreschi Architects, New Cultural Center, Pomezia, longitudinal section.

Figure 10
Lenci-Valentin, Ferentino Technological Center, Ferentino.

archaeological museum built in a historic granary. In this project, the new museum volumes intersect the existing building containing the theater. The relationship between the old and the new is characterized by the juxtaposition of the new volumes with the existing building and the use of modern materials and technologies to create a contrasting dialogue between tradition and innovation (Figure 9).

At the Ferentino Technological Center, the challenge was to renovate a former silver cutlery plant and repurpose 4,000 square meters as an innovation center. The primary design strategy included enhancing and restoring the existing structure, retaining its visibility, and creating a two-story gallery that is treated as an open-air street, with the construction of office and laboratory space. The choice of materials was essential to distinguish the existing building from the new additions while maintaining harmony between the past and the present (Figure 10).

The design procedures used in the above redevelopment projects aimed to revitalize industrial sites by solving architectural and urban issues. In-depth knowledge of the sites was necessary to determine the best design strategies. Five general procedures were identified as a means of approaching industrial heritage design. These procedures are not comprehensive but are intended to pinpoint aspects and values to focus on during the design process. A systematic approach for redevelopment projects is briefly explained below. It aims to provide a framework for identifying and analyzing critical urban and architectural issues before design decisions are made. It also establishes and orders the progression of design activities to be performed during the design process.

4. Design Procedures and Values in Redevelopment Projects

The large size and spaciousness of industrial structures make their reuse a unique, challenging, and complex project. Understanding how to identify which buildings and elements have historical–cultural value, how much to preserve, and what to use would be the most appropriate are the main critical issues to be addressed. The final evaluation and selection of an intervention can only be made after a comprehensive study and analysis of a site. A project dealing with preexistence requires complete knowledge of the immediate urban and territorial context. Regardless of the selected intervention, be it conservation, preservation, or adaptive reuse, considering a site's preexistence will be necessary, but it should not sacrifice aesthetics, new uses, or innovation. As evidence of the past, industrial heritage sites can become strategic cultural and social nodes of an area. When the valorization of heritage sites is framed within a place, the project focuses not only on the buildings but also on the area's culture. When addressing the various design challenges, five general design procedures and values needed to approach industrial heritage have been identified.

• In-depth Study and Analysis of the Abandoned Industrial Area
The architectural, technological, and structural characteristics of the buildings, the urban and natural context, the historical, social, and economic conditions, and the site's environmental characteristics should be studied. This can help identify the most appropriate type of intervention for the site and an appropriate reinterpretation of existing conditions. The main aspects to be considered in this step are data collection, documentation (archives, site inspections), and a critical evaluation.

• Development of a Strategic Urban Vision
This procedure implies the need to consider both individual buildings and the entire neighborhood. It encourages the integration and interconnection of the project area with the surrounding environment

by creating open spaces for the community. By anticipating future social spaces, the project can propose an updated urban order capable of generating new types of associations, measures, and meanings within the contemporary city. Place specificity and urban interconnections are two main aspects of this procedure.

• **Development of a Specific Design Program and a Preliminary Design Proposal**
Functional needs intertwine with a building's existing architectural and technological conditions to develop a sustainable and feasible project. Mixed-use development is one strategy that promotes social inclusion, economic activity, and the creation of facilities and public open spaces for the community. Both a cost estimate and a feasibility study are needed to ensure the success of the project. Two essential aspects of this point are the *mixité* and the development of public spaces.

• **Development of Place Identity and Place Attachment**
During the design process, a holistic approach is essential to develop place identity and place attachment, social activities, and improved quality of life. These are key ingredients to achieve the sustainable urban regeneration of industrial sites.

• **Protection and Valorization of the Historical Memory**
The reuse of industrial heritage means protecting and valorizing heritage and maintaining historical memory for the current population and future generations. Knowing our past helps to understand the richness of the present and the possibilities of the future. However, heritage valorization implies the need to establish a dialogue between preexistence and innovation, past and future.

Through the above design procedures, a thoughtful reinterpretation of specific site conditions and a sensitive assessment of values will help to achieve concrete, resilient, and sustainable results that can transform unresolved and critical areas into economically viable places that provide improved quality of life. In this sense, these procedures become a means to order, define, connect, and generate new urban spaces with social and economic activities capable of developing an extended dialogue with all or part of the city. Effects that extend beyond the site's boundary can influence the wider urban territory and lead to what may be called an urban-scale project.

5. Industrial Heritage: Systematic Design Approach

Design issues such as building and site documentation, critical evaluation, strategic urban vision, place specificity, urban interconnections, architectural programming, *mixité*, public open spaces, place identity and place attachment, sustainability, industrial heritage protection and valorization, and historical memory are primary aspects to focus on during redevelopment projects. These design issues are part of the systematic approach used in personal architectural practice and theoretical research. They follow a framework that establishes and orders the progression of design activities and use a "zoom in" approach, from the project's largest urban scale to its smallest architectural scale. However, a "zoomed out" view is also necessary to thoroughly evaluate the design process. The systematic design approach primarily used to address industrial heritage redevelopment consists of the following points:

• **Territorial Framework**
A territorial framework deals with studying and analyzing the location, geographical, and climatic characteristics, urban, environmental, social, economic, and infrastructure systems, current regional and urban plans, and significant landmarks. Presentation techniques include diagrams, images, and urban plans at scales of approximately 1:10,000 to 1:5,000.

• **Urban Framework**
An urban framework analyzes, on a smaller urban scale, aspects such as location, infrastructure systems (primary and secondary roads and pedestrian networks, public transportation), major urban and building systems (building types and heights, urban voids, landmarks, nodes, zoning), environmental systems (geographic and climatic features, wind, topography, vegetation), social and economic systems (population density, services, and facilities), and existing urban plans. Presentation techniques include diagrams, images, and urban plans at scales of approximately 1:2,000 to 1:1,000.

• **Historical Framework**
The historical study of industrial buildings and sites includes an investigation of the former uses and duration, existing uses and conditions, historical chronological development, building types, structures, and technologies, industrial production processes, existing industrial machinery and objects, urban, social, and economic situations, and environmental impacts. Presentation techniques include diagrams, images, and urban plans at scales of 1:5,000, 1:2,000, and 1:1,000.

• **Site Analysis**
A site-specific study includes the investigation of site boundaries, primary and secondary access, infrastructure systems (vehicular, pedestrian, public transportation), urban systems (building types and heights), and environmental systems (climate, wind, topography, vegetation, views). Presentation techniques include plans, elevations, and sections at scales of 1:500, 1:200, and 1:100.

• **Building Analysis**
Building analysis includes an assessment of the building types, heights, uses, and technology, cultural content (machinery, tools), building materials, structural and technological systems, and existing conditions. Presentation techniques include images, plans, elevations, and sections at scales of 1:200 and 1:100.

• **Programming**
Programming comprises predesign work, the definition of project goals and budget, cost-benefit and feasibility studies, design references, and a detailed construction schedule.

• **Preliminary Drawings**
Preliminary drawings include concept drawings, floor plans, elevations, and sections of the project proposal at scales of 1:200, 1:100, and 1:50, sketches, renderings, and a preliminary building cost estimate.

• **Design Development**
Design development is the process of creating architectural plans, elevations, and sections at scales of 1:200, 1:100, and 1:50; detailed sections at scales of 1:20 and 1:10; renderings; and a building cost estimate.

• **Construction Documents**
Construction documents include architectural, structural, mechanical, and electrical drawings; details of special systems with diagrams; plans, elevations, and sections at scales of 1:200, 1:100, and 1:50; detailed sections at scales of 1:20 and 1:10; and renderings, construction estimates, building specifications, and building maintenance manuals.

• **Bidding, Construction Administration**
Bidding and contract administration procedures are part of the final stage in which bids are evaluated, contractors are reviewed and selected, and contracts are executed and awarded.

6. Academic Research and Application Studies

The preceding design procedures and approaches have been applied in academic research and design laboratories. Two results are briefly explained here.

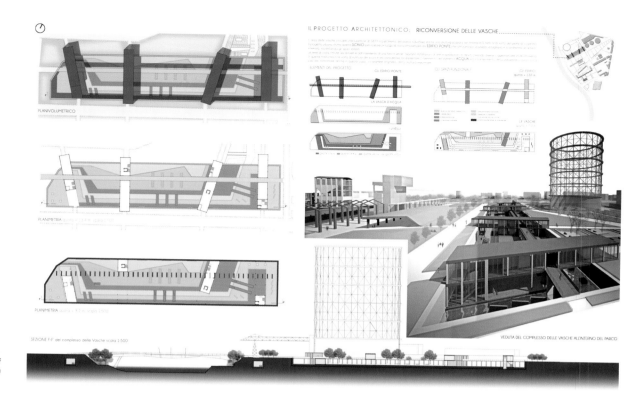

Figure 11
Academic research with A. Iacovoni, redevelopment of the former gasometer area in Rome.

6.1 Redevelopment of the Former Gasometer Area in Rome

To redevelop the former gasometer area, a vast abandoned sector of the city, a territorial and urban analysis that included studying and evaluating archived documents and site surveys was conducted. Pedestrian, vehicular, social, and environmental systems, actual site conditions, and current urban plans were also analyzed. The site is located between the Tiber River and Viale Guglielmo Marconi, the main axis connecting the historic center with EUR, the Esposizione Universale Roma, and Ostia. The historical analysis covered the area's chronological development, the existing building types, the buildings' structural and technological condition, and the site's past and present use. After analyzing the available information and the area's current infrastructure issues, a specific program and strategic urban vision were developed considering the site's criticalities and potentialities. Given the lack of open green space and the importance of the existing structures, the decision was made to create a new public urban park, restore and reuse the existing buildings, generate new cultural and social activities, develop pedestrian connections between the buildings and surrounding areas, and identify solutions to the current traffic problems. The existing concrete water tank, previously used to store coal underwater, was transformed into a community service area, with bridge buildings connecting the ground level with the new activities located on the tank's lower level. The vertical coal storage buildings, which were considered landmarks, were renovated to house services. New pedestrian paths and a bridge crossing the Tiber River were designed to form connections with the surrounding neighborhoods (Figure 11).

6.2 Redevelopment of the Former SITOCO Industrial Area in Orbetello

The abandoned former SITOCO factory, Società Interconsoriziale

Toscana Concimi, is located in a prominent natural area near the Orbetello Lagoon, close to a protected nature reserve managed by the World Wildlife Fund. The project proposed redeveloping the industrial site as a "natural science city," with botanical, exhibition, educational, research, and residential areas (Figure 12).

The main idea was to preserve the existing buildings and, thus, the historical memory of the place. Some architectural features were added by using modern materials and technologies. The design approach began with a study and urban analysis of the industrial site, with a specific investigation of the environmental and ecological context. The natural features and brownfield problems were the two main aspects explored in the investigation. The historical analysis reviewed the development of the manufacturing complex and examined the existing building and their structural and technological conditions. Available historical documentation and site investigations were used in the site analysis, functional program, and design strategies. The decision was to create a city effect by developing a mixed-use complex characterized by a central pedestrian promenade and the waterfront. The former arsenic building was transformed into the main exhibition space featuring its unique internal wooden structure (Figure 13). The project was fully developed with plans, elevations, sections, architectural details, and renderings.

7. Research Discussion

The main theme that emerges from the study of former industrial sites is the need to fully characterize the "historical, technological, social, architectural, or scientific value"[1] of the heritage site before making final decisions about a project. A well-structured analysis of the area that includes tangible and intangible

Figure 12
Academic research with M.G.
Combusti, redevelopment of
the former SITOCO industrial
area in Orbetello.

information ensures that the most appropriate architectural and urban strategies will be identified and adhered to during redevelopment.

Design procedures should use a multidisciplinary approach because the existing building types will exert unique constraints on each project. Furthermore, different projects must contend with varying degrees of adaptability and flexibility of the structures' re-functionalization, the historical–cultural value of the buildings and the surrounding neighborhood, local building codes, structural–technological characteristics, and socioeconomic and environmental conditions. Despite the variety of complex and interconnected issues, it is essential to find ethical and aesthetic solutions to problems. The multidisciplinary approach presented here is useful in the analysis and decision-making phases for the designed reinterpretation of existing architectural and urban realities.

The continuous postponement and delay of major urban regeneration projects is a current phenomenon, despite the short- and long-term benefits they would bring to the community. The reasons for delay vary from financial difficulties to the presence of brownfield sites. A growing public awareness of industrial heritage has fostered increasing interest among public and private stakeholders in the redevelopment of industrial sites; thus, it is essential to understand their cultural value and economic and social assets. The redevelopment of industrial heritage can act as a catalyst for revitalized economic and social activity at the site and in its immediate surroundings, as well as in the extended urban or territorial system.

In addition to a cost–benefit analysis, it is essential to perform a feasibility study that considers, as appropriate, any operational and fiscal costs that may compromise the long-term economic independence of the project and, therefore, its future maintenance and existence. The Venice Charter recommends the reuse (rather than abandonment) of historic buildings to prevent, at least temporarily, physical deterioration.

8. Conclusion

The design procedures and approaches briefly discussed in this work are important for developing a regeneration project that uniquely responds to a particular place. The intention of these procedures and approaches is to develop a site as a well-integrated and essential part of a more extensive urban system.

While limiting urban sprawl, the reuse of former industrial sites can create mixed-use developments, public squares, and social activities that not only increase the economic value of a place but also augment its identity and character. In this way, these projects can improve the quality of life, image, and aesthetics of the site and the wider neighborhood.

A full understanding of the historical, social, environmental, and cultural values of a site also helps in the development of unique design strategies that can achieve resilient, flexible, and sustainable design solutions and reinterpretations. This process is necessary not only for the future of the industrial heritage but also for the future of the community. Indeed, industrial heritage offers the opportunity to nurture historical memory and people's attachment to place—values that are important for cities' resilience.

Educators are responsible for preparing future generations to work with abandoned, challenging, and unresolved industrial areas. Dealing with industrial heritage, or any heritage, involves extending one's knowledge beyond the usual design processes to include the

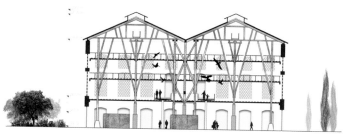

Figure 13
Redevelopment of the
former SITOCO industrial
area in Orbetello.

multilayered history presented at a project's architectural, urban, and territorial levels. Thoroughly understanding the existing urban fabric, site, environment, and building history leads to insights into the specific architectural needs of a project.

The valorization of industrial heritage remains a delicate issue that is not always resolved or addressed in many countries. The role of institutions and communities is crucial when discussing the future of degraded industrial areas and their potential reuse.

There is a growing opportunity to think of these areas not as negative places to avoid but as sustainable resources. Given the deep economic, social, and environmental complexity of industrial heritage projects, it is necessary to pursue public discourse on sustainable strategies to deal with them at the national and global levels. These sites should not simply be capped with concrete, which would destroy large natural areas, because these decisions also affect the larger ecological system.

Industrial archaeology, industrial heritage, and industrial landscape have become central environmental and urban issues. With deindustrialization, we have witnessed the abandonment, environmental degradation, depopulation, and marginalization of entire industrial areas. A deeper reflection on these "suspended" sites is essential for the future transformation of contemporary cities.

Acknowledgments

The author acknowledges the research team, which includes Alessia Iacovoni for the redevelopment of the gasometer area and Maria G. Combusti for the redevelopment of the former SITOCO industrial area.

Notes
1. TICCIH, "The Nizhny Tagil Charter 2003."

Figures
Figure 1: Luciano Cupelloni, MACRO Future Testaccio Museum of Contemporary Art, Rome, 2002–07 (author's photo).
Figure 2: Centrale Montemartini Museum, Rome, 2005 (author's photo).
Figure 3: Former gasometer, Rome, 1935–37 (author's photo).
Figure 4: Zaha Hadid, MAXXI, Museum of 21st Century Arts, Rome, 2010 (author's photo).
Figure 5: Odile Decq, MACRO, the Museum of Contemporary Art of Rome, 2010 (author's photo).
Figure 6: Lenci-Valentin, study for the new master plan of Rome in 2000 (author's photo).
Figure 7: Lenci-Valentin, proposal for the urban redevelopment of Borghetto Flaminio, Rome (author's photo).
Figure 8: Lenci-Valentin, proposal for the transformation and urban renewal of San Lorenzo, Rome (author's photo).
Figure 9: Petreschi Architects, New Cultural Center, Pomezia, longitudinal section (author's photo).
Figure 10: Lenci-Valentin, Ferentino Technological Center, Ferentino (author's photo).
Figure 11: Academic research with A. Iacovoni, redevelopment of the former gasometer area in Rome (author's photo).
Figure 12: Academic research with M.G. Combusti, redevelopment of the former SITOCO industrial area in Orbetello (author's photo).
Figure 13: Redevelopment of the former SITOCO industrial area in Orbetello (author's photo).

Urban Regeneration in a Capital City: The Case of Islamabad, Pakistan

Samra Mohsin KHAN, COMSATS University Islamabad, Pakistan

Abstract

Constructed in the 1960s as the first capital of the new nation of Pakistan, Islamabad was designed by Doxiadis and reflects his vision and philosophy of a dynapolis and city of the future. Its planning, sectors, and architecture fully reflect the International Style, signifying an era of modernity that breaks with history through the creation of new urban fabric and architecture. This paper traces the city's development over fifty years, showing how the original master plan, conceived in the 1960s, failed to address cultural factors that have shaped South Asian cities. Multiple developmental failures have constrained efforts to expand the master plan's guidelines to resolve issues that are contributing to Islamabad's physical and social decline. Current international urban regeneration projects are explored to identify sustainable urban regeneration approaches that could revitalize Islamabad's urban space, while preserving its history and reputation as a model modern city through revisions of the original master plan.

Author's Information
Samra Mohsin KHAN: Samramkhan@gmail.com

Keywords

Islamabad, urban regeneration, master plan.

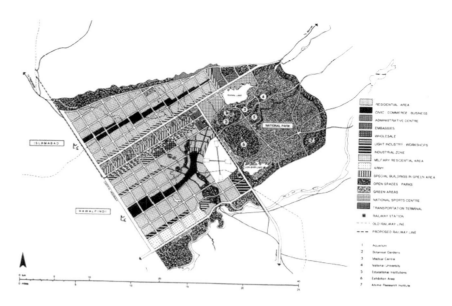

Figure 1
Final master plan for
Islamabad, 1960.

Figure 2
Vedat Dalokay, Faisal
Mosque, Islamabad, 2011.

Figure 3
WS Atkins, the Centaurus,
Islamabad, 2019.

1. Introduction: Islamabad's Master Plan as Envisioned by C.A. Doxiadis[1]

Islamabad is one of the few capital cities designed in the twentieth century that reflect the combined use of two concepts: ekistics and dynapolis in its planning.[2] Doxiadis expressed this vision within innumerable conceptual maps, notes, and sketches that he subsequently translated into physical form. The capital city's identity was forged through its innovative planning and many iconic buildings designed by star architects of the modern era, notably Louis Kahn, Arne Jacobsen, Gio Ponti, Sir Robert Matthew, Edward Durell Stone, Kenzo Tange, Harvey Foster, Denis Brigden, and Vedat Dalokay. (Figure 1 and 2)

Doxiadis designed the city's housing sectors as self-contained communities with schools, places of worship, shops, and parks and playgrounds. Overall, the wide avenues and suburban-oriented planning were designed to be automobile-friendly. While Islamabad accommodated government functionaries along with the urban upper and middle classes, it failed to respond to the residential needs of the hundreds of thousands of workers who were needed to service the city. In the 1960s and 1970s, Islamabad was considered a dream city that appeared to be "planned for the future and built for the present."[3] The capital became a magnet, attracting a massive influx of migrants from the city's hinterlands and rural areas. The original master plan was visionary in its conception and scale and was expected to undergo careful reviews every twenty years to preserve the city's salient features while planning its future expansion.[4] The Capital Development Authority (CDA) attempted to revise Islamabad's master plan every few years and address emerging needs for infrastructure and amenities. However, no comprehensive review of the master

13

Figure 4
Greenwich, London, 2018.

Figure 5
King's Cross and St. Pancras
stations, London, 2018.

plan has been undertaken over the last fifty years, resulting in a lack of feasible large-scale development projects. Instead, approximately 40 minor revisions were made to the master plan, leading to selective and fragmented developmental work. Consequently, the city's rapid urbanization and resultant issues were never addressed holistically.

1.1 Islamabad's Contemporary Urbanization Issues

The rapidly expanding city of Islamabad currently has a population of over 1.147 million.[5] A growing population, poor planning, and the lack of enforcement of legislation has prompted the development of slums in the city as well as inadequate waste management, water issues, lack of public transport, environmental issues, declining infrastructure, and the mushrooming of unauthorized constructions, particularly in the capital's periurban areas.[6] Small retail centers in the sectors and subsectors have been replaced by new types of retail outlets like supermarkets and shopping malls. Consequently, parking in the city has become a major problem, and congestion and traffic jams are an everyday phenomenon. Long neglected and inadequately maintained, the city's original parks and green belts have become degraded, or else they have been encroached or converted into local garbage dumps.[7] The increase in car ownership has prompted the widening of roads and the construction of massive overhead carriageways and underpasses that have entailed extensive tree felling and ecological destruction, adding to rising air pollution levels. (Figure 3)

Islamabad's car-oriented planning impedes pedestrians' easy and rapid movement in the city centers. The clusters of buildings in the city have been planned around vast open spaces that are mostly used as surface parking lots, making them esthetically unappealing and contributing to the urban heat island effect. The existing design, which entails few footpaths and trees, and

therefore inadequate shade, provides very limited opportunities for pedestrians to walk around the city. To develop a comprehensive understanding of the challenges that the city is facing and to develop proposals for its sustainable regeneration, it is necessary to grasp the aims of urban regeneration, the mechanisms entailed in the regeneration process, and its physical, economic, and social consequences. Accordingly, a study that examines the experiences of other cities across the world that have already implemented urban regeneration projects can elucidate the processes that were followed and demonstrate successful outcomes.

2. Urban Regeneration

Urban regeneration is usually associated with historical cities that were once the grand urban centers of great civilizations. Most of these cities, such as London, Rome, and Paris and, within Asia, Shanghai, Delhi, and Lahore are old major cities that have existed for hundreds of years. They have developed and grown over time from smaller-sized towns to their current forms as overgrown urban juggernauts. Their issues span centuries, and urban expansion and development that occurred outside the boundaries of the original cities usually comprised organic and unplanned sprawls that were later given some organizational coherence by urban designers and planners. Historically, "urban regeneration" has been perceived as a way of revitalizing declining cities in North America and Western Europe, where urban dynamics have been affected by economic restructuring, the decline of traditional industries, and the decaying urban fabric of inner cities.[8]

Regeneration projects implemented in city centers generally incur high costs for both the public sector and private developers, with the

Figure 6
King's Cross, London, 2018.

residents of these areas at times being forced to move elsewhere. The pressure on developers and the need to ensure financial viability may lead to the demolition of historical buildings and their replacement by high-rise buildings or high-density complexes. The compulsion to acquire maximal commercial profits generally takes precedence over initiatives to preserve old buildings.[9] This situation accounts for the lack of character displayed by many of the new, large-scale redevelopment and gentrification projects that have entailed the removal of indigenous social and cultural features. Therefore, a context-sensitive approach that entails an understanding of local issues and the proposal of sustainable solutions is essential to ensure the successful outcome of a regeneration project.[10]

To be successful, urban regeneration should be aimed at solving urban problems and finding sustainable solutions that help to improve the physical, social, environmental, and economic features of cities and should include clear and measurable objectives, an analysis of local conditions, and participation of all stakeholders.[11] Social sustainability and social equity are critical dimensions of sustainable urban development. Accordingly, all citizens should have equal access to resources and to the social, economic, recreational, and other facilities that a city offers.[12] However, Islamabad's linear and car-centric design caters primarily to rich, car-owning citizens while marginalizing the non-car-owning populace. The lack of adequate public transport, pedestrian areas, footpaths, sheltered pathways, and public squares makes the city less accessible to ordinary citizens.

The development of a sustainable regeneration proposal for Islamabad, which requires a comprehensive understanding of the challenges that the city faces, necessitates a broad study to extract lessons derived from the experiences of cities worldwide that have already implemented successful urban regeneration projects. These examples can elucidate the processes followed and reveal successful outcomes. Moreover, an understanding of the aims and mechanisms of regeneration projects and their physical, economic, and social consequences is required.

2.1 London

Eighty percent of the UK's population comprises urban dwellers. These demographic conditions sparked the idea of restricting outward urban growth and promoting efficient recycling of resources in the 1920s. The Town & Country Planning Act of 1947 underscored the multiple objectives of designing and retaining "green belts" "to check the unrestricted sprawl of large built-up areas" and "to assist in urban regeneration, by encouraging the recycling of derelict and other urban land."[13] (Figure 4)

Subsequently, the development of a compact city emerged as the preferred response for achieving the goal of sustainable development by providing "dense and proximate development patterns, built-up areas linked by public transport systems, walkability and accessibility to local services and jobs."[14] Two regeneration projects implemented in London are presented below to demonstrate the application of two contrasting approaches for implementing urban regeneration.

2.1.1 Redevelopment of King's Cross Station

The redevelopment of the area around King's Cross and St. Pancras railway stations in central London is illustrative of a compact, mixed-use regeneration project. (Figure 5 and 6) The project is one of the largest regeneration schemes in Europe, entailing the transformation of sixty-seven acres of industrial wasteland into

15

Figure 7
University of the Arts London,
2018.

Figure 8
Laurie Olin, Lewis Cubitt
Square, King's Cross, 2018.

a mixed-use, vibrant neighborhood with homes, shops, offices, galleries, bars, restaurants, schools, and a university. There are over 1,000 choreographed fountains in the Granary Square. South of this square are large steps leading down to the canal, which is a popular lunch spot. The square is home to many different restaurants and to the relocated University of the Arts London building, which adds a cultural dimension to the site (Figure 7). Parks and gardens have been incorporated into the design of this area, including the Lewis Cubitt Square (Figure 8) and the Lewis Cubitt Park, the Gasholder Park, and smaller urban gardens where residents of these areas are permitted to grow their own vegetables. The Gasholder Park is a small green space with seating and optical sculptures installed inside the largest gasholders. (Figure 9 and 10)

The King's Cross master plan envisions the construction of 2,000 new homes, fifty new buildings, twenty new streets, and twenty-six acres of parks, squares, gardens, and open space, each designed by a different planner.[15] The Regent's Canal was incorporated into the plan as an important design feature, with strategically placed public seating and buildings. The site entails mixed land use, of which forty-seven percent comprises office space, ten percent is allocated for educational facilities, twenty-five percent for residential homes, and the remaining portion for hotels, retail and leisure/other facilities.[16] The mega project will help to recycle railway wasteland, incorporate the extensive development of public squares, parks and gardens as well as green roofs and walls, and most importantly the restoration of historic buildings of the area.

Although the development project highlights the preservation of listed buildings, it also prioritizes "social inclusion" by addressing the needs of the local community. However, the project has been criticized because of the limited provision for affordable housing, with most of the residential stock comprising luxury apartments and penthouses. Consequently, the area will become more gentrified, leading to a rapid escalation in property prices, and the few youth clubs or meeting places in the area will end up becoming private/exclusive environments.

The project was awarded high environmental sustainability ratings, as it includes a low carbon strategy, aimed at achieving a sixty percent reduction in carbon between 2000 and 2050. Moreover, the master plan provides for elements of community infrastructure across the site and entails a strong social vision. This project to redevelop the King's Cross area of central London offers pertinent lessons relating to the quest to balance urban compaction with livability, while addressing the specific challenges presented by a large global city.

2.1.2 Ilford Town Centre

In contemporary London, a scheme known as the "Healthy Streets Approach" is underway, aimed at reinventing the use of streets to make them healthy, safe, and welcoming for everyone. The scheme foregrounds people and their health in decisions about how public space is to be designed, managed, and used by the public.[17] It focuses on creating designs for public space, both large and small, for diverse uses to improve conditions for walking, cycling, and traffic reduction. An important component of the scheme entails developing proposals through engagements with local communities. The program budget amounts to £139 million to be spent over five financial years (2018/19–2022/23). (Figure 11 and 12)

Ilford is projected to experience a twenty-nine percent increase in growth that can be attributed to the inclusion of Ilford Station on the Elizabeth Line. To accommodate this anticipated growth, the Borough

Figure 9
Gasholder Park, King's Cross,
London, 2018.

Figure 10
Gasholder Complex, King's
Cross, London, 2018.

of Redbridge commissioned a multi-phase public scheme to develop new, high-quality, and accessible public spaces that would transform Ilford High Road. The project was awarded to the firm METIS Consultants Ltd., London. The firm undertook a comprehensive and lengthy process of engaging all stakeholders and seeking their inputs on the final design as well as adopted a strong advocacy campaign. The scheme includes hard and soft landscaping, the provision of seating and informal play areas, and the improvement of pedestrian and cycle routes.[18] Public art, new trees, and pocket parks will be part of a multimillion-pound street improvement project for Ilford Town Centre in which the enhanced streetscape will be aligned with the Ilford manifesto of "Ilford for all."

The scheme aims to contribute to improving London's urban fabric and promoting lifestyles that foster residents' physical and mental wellbeing within their city. The designers have formulated a proposal that provides flexibility and opportunities for integration. Ilford Town Centre not only encompasses shopping, local markets, and play areas for families but it also includes elements of public art, installations, spaces for holding public events, and outdoor food markets to attract younger residents. Thus, the present social setting is intended to encourage families to visit their high street while also introducing a range of activities to attract younger individuals.[19] The design will be merged into the existing urban fabric, benefiting current users in addition to fostering future development.

Urban regeneration projects generally foreground a preservation paradigm (sustaining historical buildings and the urban fabric) in conjunction with a developmental paradigm (retail, residences, and recreation), turning the historical and often decaying urban fabric into a holistic and economically viable project. In Europe this approach has evolved over a century and a half, whereas in China,

its progress has been rapid, unfolding over the last three decades. The presentation of heritage as tourist attractions along with economic development, and wealth creation in China have been part of a wider strategy of branding culture as a key economic commodity.[20]

2.2 The Regeneration of Xintiandi in Shanghai

Shanghai has been an international metropolis within East Asia commencing from the nineteenth century. The city flourished, emerging as a primary commercial and financial hub in the Asia-Pacific region in the 1930s. However, infrastructure and housing development were impeded by budgetary constraints, leading to the deterioration of Shanghai's physical environment. Consequently, by the end of the 1970s, the city had become congested given its large population and underdeveloped infrastructure. In the 1990s, economic reforms enabled the initiation of intense redevelopment schemes, especially in the Pudong New District, once again attracting financing and foreign investments[21] and reestablishing the city's status as a hub for international trade and finance. (Figure 13 and 14)

Today, Shanghai is a bustling and wealthy city where livability and life quality have been upgraded for the city's residents and safe and well-designed public spaces have been developed. The city boasts adequate housing stock, an excellent public transport network, enhanced social and recreational opportunities, a significant expansion of public and green spaces, and a series of successful regeneration projects that have been implemented in the city.

The redevelopment of the old *shikumen* residences in Xintiandi, which is one of the busiest areas in Shanghai, has resulted in the area's transformation into one of the city's commercial and cultural centers. (Figure 15 and 16) The redevelopment project

Figure 11
METIS Consultants Ltd., Ilford
High Road, London, 2020.

Figure 12
METIS Consultants Ltd., Ilford
High Road, London, 2020.

Figure 13
Model of Shanghai city, 2019.

Figure 14
Wujiaochang, Shanghai,
2019.

Figure 15
Shanghai's *lilong* from the 1920s, 2019.

Figure 16
Lilong, Shanghai, 2019.

was launched in 1998 following the repaving of alleys in this area. *Shikumen* residences were painted and transformed into retail stores for selling artwork, clothing etc., or into tea houses, while suitable factories were converted into art studios.[22] (Figure 17 and 18) In 2003, the Urban Land Institute (ULI) Award for Excellence and the American Institute of Architecture (AIA) Citation for Heritage were conferred on the Xintiandi Project in light of its achievements. In Xintiandi's adaptive reuse project, the residential function of the *lilongs* was converted to mixed commercial use.[23] Whereas the interiors of the *lilong* buildings were redesigned and transformed into coffee shops, restaurants, and bars, their original façades were cleaned and preserved. The main alleys were paved with granite, and the secondary alleys were paved with bricks obtained from demolished buildings. (Figure 19 and 20)

The downside of the development project was the relocation of *lilong* residents and their exclusion from the planning process. The preservation of *lilong* residences has clearly been worthwhile in terms of creating an attractive tourist destination that boosts the economy. However, the removal of the original residents from the area has led to the loss of its living culture and social fabric, with just its "shell" preserved in the form of the *lilong* buildings. The development of upscale commercial and residential areas inevitably led to the area's gentrification. The lessons gained from the Xintiandi initiative indicate the need to balance economic benefits with an improved lifestyle for residents and the creation of more jobs for them within future regeneration initiatives.

2.3 The Regeneration of Pingjiang Road and Its Surroundings in Suzhou
The historical Pingjiang Road is located in the old urban section of the city of Suzhou, approximately 100 kilometers west of Shanghai. This road is about 1.5 kilometers long, with historical buildings lining both

sides of the road along the central canal. Pingjiang Road and the Pingjiang district in which it is located have recently undergone socially sensitive regeneration that has transformed them into a tourist center within the city. Heritage preservation and regeneration efforts in this area have emphasized both its tangible and intangible qualities. Existing social networks have been maintained during the project, and residents continue to live and work within the district, preserving their traditional culture and customs.[24]

This district exemplifies regeneration that preserves traditional Chinese architecture and the living customs of a traditional society. The project has focused on the restoration and functionality of historical houses, the adaptive reuse of some buildings into cafes and tea rooms, cleaning shop façades as well as the canal and roads, installing street furniture, and preserving the landscape and the environment. The regeneration initiative not only restored buildings, waterways, and bridges, but it also revitalized intangible social and cultural dimensions, such as festivals, and improved living conditions for residents, while supporting local businesses. (Figure 21 and 22)

A stroll through the area provides a vista of modern shops with glimpses into the traditional lifestyle of the Chinese people. The regeneration of Pingjiang Road offers a model for sustainably merging improved living standards for the residents with tourism and cultural preservation with economic benefits for all stakeholders.

3. Islamabad: Regeneration of Saidpur Village

Saidpur is a small and historical 200-year-old village located in the foothills of the Margalla Hills about 500 meters from F-6, an upmarket residential sector in Islamabad. The village has about 15,000 residents and comprises 1,500 households. In 2006,

Figure 17
Xintiandi, Shanghai, 2019.

equipped with a budget of nearly 400 million rupees (approximately 3.5 million US dollars), the CDA undertook a regeneration initiative to transform Saidpur into a "model tourist village."[25] The renovation project, which was mostly completed by 2008, was aimed at preserving Saidpur as a cultural heritage site that would attract tourists and local residents of Islamabad. It included a plan to develop a major arts and crafts village in the vicinity to provide residential and shop spaces for the artisans and display booths. The redeveloped village center comprised a combination of restaurants, cafes, handicraft and souvenir shops, a museum, and an art gallery. (Figure 23 and 24)

3.1 The Regeneration Process
The CDA promised the residents of Saidpur that they would benefit from the renovation project, which would provide them with an improved water supply, a gas supply, proper drainage, and a garbage removal system. Moreover, it would create employment opportunities for the locals and support economic development in the village. The CDA acquired land in the village center, which was to be developed by demolishing the houses of villagers after promising to compensate and relocate them. The scope of work for this project encompassed the restoration and conservation of heritage buildings, the acquisition and demolition of residences located along the main road, the development of dam/water retaining structures, landscaping and horticulture, the provision of a water supply, sewage works, the construction of pedestrian and vehicle bridges, road infrastructure, electric work, and other civil works.[26]

The pre-existing historical structures at the site were a Hindu mandir (temple), a second Sikh temple, and a dharamshala (orphanage) located in an elevated site in the center of the village. The two temples were repaired and repainted while the two-story dharamshala was repaired and converted into a museum for the city of Islamabad. The re-adaptive use of the dharamshala as a museum and gallery space introduced a tourist magnet into the village, forging a strong village identity. (Figure 25 and 26)

3.2 The Successes and Failures of the Regeneration Project
In spite of the pledges that the CDA made to the community, little sustainable development took place. The economy of the village did initially benefit from this regeneration initiative because local residents gained employment in the building trade and by hiring artisans. Local shopkeepers also gained some business catering to requirements associated with the increased construction activities in the village. However, in the long run, outsiders who own and manage the restaurants that opened along the main village street captured most of the benefits.[27]

Social exclusion was an unintended outcome of the Saidpur regeneration project, given the establishment of many fashionable eateries that are unaffordable for the locals. The environmental improvements envisaged for the site were not realized, as no concrete steps were undertaken for the removal of garbage from the village. Consequently, garbage continued to be dumped in the natural canal that runs through the village. (Figure 27) Moreover, economic benefits have not been transferred to the village residents.

While attempting to revive economic activities in the area and to balance the CDA's economic inputs, the project did not account for the losses suffered by displaced residents. Moreover, the gentrification of the village's central public space associated with the new commercial and retail enterprises was not anticipated. Given the lack of consideration of social, cultural, and ecological issues, little

Figure 18
Xintiandi, Shanghai, 2019.

Figure 19
Seating areas, Xintiandi,
Shanghai, 2019.

Figure 20
Courts with cafes and shops,
Xintiandi, Shanghai, 2019.

Figure 21
Fashionable boutiques targeting tourists, Pingjiang Road, Suzhou, 2019.

Figure 22
A tea shop, Pingjiang Road, Suzhou, 2019.

has changed in the lives of the residents.

In the absence of any advocacy campaign, the local community has experienced social as well as physical marginalization. Evidently, it is important not only to consider the influence of market forces within regeneration projects but also to explore how the social and cultural fabric of the area can be preserved by ensuring residents' participation as stakeholders of the proposed development initiative.

4. Urban Regeneration Strategies for Islamabad

The following conclusions can be drawn based on lessons derived from the above examples of regeneration projects implemented in various city centers across the world.

Approaches for regenerating city centers that do not balance residents' interests with tourism and the high costs incurred for the government and private developers are less successful. Economic feasibility is the driving force behind the destruction of historical buildings and the removal of residents to make way for expensive residential and commercial development.

Regeneration should be sensitive to the existing social and cultural nuances of an area. Moreover, it should be aimed at creating a sustainable environment that promotes human interactions within spaces designed for socialization, namely streets, squares, and parks along with facilities for sitting, walking, and cycling.

The creation of mixed-use neighborhoods that entail overlapping residential, recreational, and commercial areas and the promotion of walking and reduced dependence on cars and transport is a third recommendation emerging from this study. The conversion of the existing historical fabric into mixed commercial areas for promoting local crafts and benefiting shop owners as well as

larger retail complexes is also advantageous.

The most effective urban regeneration programs have been premised on the conviction that historical continuity requires the preservation of both the physical and the social/cultural environment. Projects that have ignored the issues faced by the original residents of the project sites have lacked historical and cultural relevance and have only benefited developers and a small group of upmarket consumers. Thus, regeneration initiatives should consider not only market forces and the interests of end-users but should also address two other important dimensions of urban development: social needs and environmental sustainability.

Islamabad was conceived as a linear city with a series of core areas that would develop parabolically like the hypothetical dynapolis. (Figure 28) However, fifty years after Islamabad's establishment, it has become evident that the city's linear expansion has encroached on its hinterlands in addition to having irreparable environmentally adverse impacts through the increased use of automobiles as a mode of transport within the city. These ongoing impacts, which include air pollution, resource depletion, and damage to the city's ecologies, are extensive. Therefore, Islamabad will need to retain its boundaries while increasing its existing urban density, which will decrease its dependence on automobiles. The following key urban regeneration strategies will need to be implemented to ensure its sustainable development:

(1) Denser, mixed-use projects that provide residential, recreational, cultural, retail, and office spaces should be designed, and walkability and a healthy and active lifestyle for residents should be promoted.

(2) Interventions that transform at least fifty percent of the city's central area into pedestrian zones, thereby encouraging increased

Figure 25
Polo Lounge, Saidpur Village, Islamabad, 2012.

physical activities through walking and cycling, while also managing traffic within this area are required. Streets should be designed as meaningful and open public spaces that are welcoming and appealing as well as safe, nurturing, and fully accessible to all so that people want to spend time in these spaces. (Figure 29) Such designs should include landscaping, the installation of appropriate street furniture, the provision of diverse spaces for children, the elderly, and women, and spaces for indigenous markets, activities, and events.

(3) Islamabad's heritage should be preserved through urban redevelopment projects that encompass modern architecture, the city's heritage areas, public spaces, and parks. Sustainable practices should be deployed to maintain a balance between residents' needs, developers' profits, and the upgradation of the area. Moreover, the adaptive reuse of older buildings should be encouraged to generate tourism, create jobs, and improve the area's economy.

(4) Accessibility and linkages throughout the city should be improved through the development of efficient mass transit systems, notably a new underground metro system, a light rail system, and buses.

(5) New planning guidelines should be drawn up that not only consider the needs of upmarket areas in the city but also address issues in the city's slums and relevant infrastructure required to improve the quality of life of residents in these areas. Further, areas designated as parks and green areas in the original master plan should be retained as the lungs of the city, thereby preserving Islamabad's ecological environment and ensuring its continuation into the next century.

5. Conclusion

Urban regeneration entails strategic development and investment to

Figure 23
Capital Development Authority, Saidpur Village, Islamabad, 2012.

Figure 24
Capital Development Authority, Restored historical structures, Saidpur Village, Islamabad, 2012.

Figure 26
New restaurants, Saidpur Village, Islamabad, 2012.

Figure 27
Decaying core, Saidpur Village, Islamabad, 2012.

create a high-quality urban environment. The future development of Islamabad will require the reinterpretation and expansion of Doxiadis' original design and principles in the city's master plan. The introduction of a new strategic development and regeneration approach that fosters an improved public domain in Islamabad is required. Accordingly, the application of the compact city model for the city centers can foster more vibrant, inclusive, and livable spaces and a sustainable urban form. These developments will result in ensuring the accessibility of all activities close to each other and allow mobility without cars. The city needs to grow around centers of social and commercial activity that are located at nodes accessible by public transport. These nodes can provide the focal points around which neighborhoods can develop. Priorities include designing clean transport systems and accelerating their use as well as rebalancing the use of streets in favor of pedestrians and communities. The future of Islamabad lies in the development of more compact versions of its traditional sectors that are higher in density and incorporate more mixed-use and high-rise developments than existing ones, thereby enabling residents to enjoy a diversity of overlapping private and public activities. Regeneration projects can utilize derelict and underused areas of the city and adopt an integrated approach that responds to social, economic and environmental issues faced in the city.

The transformation and regeneration of Islamabad will hinge on relevant studies, research, legislative proposals, policy formulation, and residents' inputs, which can all contribute to the regeneration of the city and its environment. Urban regeneration provides an opportunity to solve Islamabad's problems through activities that include rehabilitating historical areas, modernizing urban infrastructure, improving living conditions in residential sectors as well as public

Figure 28
A view of the city from
Shakarparian Hill, Islamabad,
2012.

transport networks, encouraging walkability, redeveloping public spaces (squares and parks), and installing urban furniture. The original concept of this sprawling city with low urban density must be revised to rationalize continual widening of roads, building overhead bridges and underpasses, and destroying green areas, parks, and the city's ecologies. Sustainable regeneration of this complex urban environment can only be achieved through cooperative efforts and collaboration among government institutions, academics, urbanists, environmental associations, developers, and, most importantly, communities.

Acknowledgments

I am grateful to Prof. Zhang Jianlong at the Department of Architecture, College of Architecture and Urban Planning, Tongji University for providing me with the opportunity to travel to Shanghai in May 2019 and for accompanying me on visits to various regeneration projects in Shanghai and Suzhou. I would also like to thank design architect and urban designer Zara Khan at METIS Consultants, London, for providing me with information about her design for the Ilford City Centre project. However, because the project was still under construction, the proposal could not be shared here in its entirety. All images of London have been provided courtesy of Zara Khan. I would further like to express my gratitude for funds provided by COMSAT University Islamabad to conduct studies in the village of Saidpur in Islamabad.

His works include regional, city, and community planning and dwelling designs. Doxiadis developed some revolutionary ideas on how human settlements should be designed, notably his concepts of ekistics and the dynapolis for designing a city of the future. He also designed the city of Sadr in Iraq, which was built in 1959. Moreover, he prepared a master plan and program for the development of the entire state of Guanabara, encompassing Brazil's former federal capital district and including the city of Rio de Janeiro in 1964.

2. Islamabad the Capital of Pakistan. Accessed December 11th. https://www.doxiadis.org/Downloads/Islamabad_project_publ.pdf.

3. I. M. Frantzekakis, "Islamabad, a Town Planning Example for a Sustainable City," Sustainable Development and Planning IV1 (2009): 175.

4. Ahmed-Zaib K. Mahsud, "Doxiadis' Legacy of Urban Design: Adjusting and Amending the Modern," Ekistics (2006): 241-263.

5. Islamabad, Pakistan, Population. Accessed December 12th, 2020. https://populationstat.com/pakistan/islamabad.

6. Qaisar Khalid Mahmood and Hassan Raza, "Peri-urbanisation: A New Challenge for Islamabad," Paper presented at the 3rd Pakistan Urban Forum, Lahore, Pakistan, December 2015.

7. H. Shaikh and I. Nabi. "The Six Biggest Challenges Facing Pakistan's Urban Future," The International Growth Center, Accessed December 15th, 2020. https://www.theigc.org/blog/the-six-biggest-challenges-facing-pakistans-urban-future/.

8. Charles Fraser, "Change in the European Industrial City," Urban Regeneration in Europe (2003): 17-33.

9. Y. Chen, "Regeneration and Sustainable Development in the Transformation of Shanghai," WIT Transactions on Ecology and the Environment 81 (2005): 235-244.

10. Michael Pacione. "Sustainable Urban Development in the UK: Rhetoric or Reality?" Geography 92, no. 3 (2007): 248-65. Accessed December 8, 2020. http://www.jstor.org/stable/40574338.

11. Peter Roberts, "The Evolution, Definition and Purpose of Urban Regeneration," Urban Regeneration (2000): 9-36.

12. Mike Jenks and Nicola Dempsey, "The Language and Meaning of Density," Future Forms and Design for Sustainable Cities (2005): 287-309.

13. Nick Gallent et al., Introduction to Rural Planning. Routledge, 2008.

14. Nicola Dempsey, Caroline Brown, and Glen Bramley, "The Key to Sustainable Urban Development in UK Cities? The Influence of Density on Social Sustainability," Progress in Planning 77, no. 3 (2012): 89-141.

15. Landscape, King's Cross. Accessed December 11th, 2020. https://www.kingscross.co.uk/media/Kings-Cross-Landscape-Brochure-vlr.pdf.

16. Marco Adelfio, Iqbal Hamiduddin, and Elke Miedema, "London's King's Cross Redevelopment: A Compact, Resource Efficient and 'Liveable' Global City Model for an Era of Climate Emergency?" Urban Research & Practice (2020): 1-21.

17. High Street for All - Report, Greater London Authority, September 2017. Accessed December 12th 2020. https://www.london.gov.uk/sites/default/files/high_streets_for_all_report_web_final.pdf.

18. Ilford Town Centre-Phase II. Accessed December 16th 2020. https://metisconsultants.

Notes

1. Constantinos A. Doxiadis (May 14, 1913–June 28, 1975) was an urban planner and architect whose major commissioned work was the design of Islamabad as the new capital of Pakistan in the 1960s.

Figure 29
Arif Masood, the National
Monument, Islamabad, 2017.

co.uk/case-studies/ilford-town-centre-phase-ii/.
19. Zara Khan (Design Architect, METIS, London.) Interview with Samra M. Khan on January 8th, 2021.
20. John Pendlebury and Heleni Porfyriou, "Heritage, Urban Regeneration and Place-Making," Journal of Urban Design 22, no. 4 (2017): 429-432.
21. Victoria Wu, "The Pudong Development Zone and China's Economic Reforms," Planning Perspectives 13, no. 2 (1998): 133-165. Also see Yawei Chen, "Financialising Urban Redevelopment: Transforming Shanghai's Waterfront," Land Use Policy (2020): 105126.
22. You-ren Yang and Chih-hui Chang, "An Urban Regeneration Regime in China: A Case Study of Urban Redevelopment in Shanghai's Taipingqiao Area," Urban Studies 44, no. 9 (2007): 1809-1826. Also see Harry Den Hartog, "Searching for a New Identity in a Rapidly Transforming Urban Landscape."
23. Lilong is the traditional Chinese house which is like a terraced English house with a courtyard inside.
24. Balancing the Old and New — Lee Kuan Yew World City Prize. Accessed on December 20th, 2020. https://www.leekuanyewworldcityprize.gov.sg/resources/case-studies/pingjiang-historic-district/. Also see Jing Xie and Tim Heath, "Conservation and Revitalization of Historic Streets in China: Pingjiang Street, Suzhou," Journal of Urban Design 22, no. 4 (2017): 455-476.
25. Muhammad Shoaib Khan, "Tourism Influencing Occupational Structures in Saidpur Village, Islamabad," Accessed on December 20th, 2020. https://www.researchgate.net/publication/265396843_Tourism_Influencing_Occupational_Structures_in_Saidpur_Village_Islamabad.
26. Naeem Khan (Architect CDA) Interview with Samra M. Khan in 2010.
27. Samra M. Khan, "Revitalizing Historic Areas; Lessons from the Renovation of Saidpur Village, Islamabad," in Sixth Seminar on Urban and Regional Planning, Seminar Proceedings (2011), 150-162.

Figures

Three Kinds of Ideal Place: A Historical Review of Shanghai's "Columbia Circle"

FENG Lu, Wuyang Architecture, China
FENG Li, Shanghai Jiao Tong University, China
WU Jiao, Tongji University, China

Authors' Information
FENG Lu: owa.china@qq.com;
FENG Li: 156150928@qq.com;
WU Jiao: a_deserts59@163.com.

Abstract

Columbia Circle is located on the west side of downtown Shanghai. It is a historic area that has been transformed over the past 100 years and has recently been renewed to become a commercial park and urban public space. It was transformed over three main periods, which correspond with three different conceptions of the ideal place. In the beginning, the construction of the Columbia Country Club and the real estate development of the 1920s reflect the dream of living in a garden suburb. From the 1950s to 2016, the place was used by the Shanghai Institute of Biological Products (SIBP) and was transformed into a *danwei*, an ideal form of collectivism in China. Lastly, after the institute moved away, the place was transformed and reopened in 2018. As an urban public space, it aims to join in the urban regeneration of Shanghai, in which community regeneration has recently come to be regarded as providing the core energy for better urban living in the future.

Keywords

Urban regeneration, garden suburb, *danwei*, community.

1. Introduction

Urban regeneration is not only the focus of contemporary Shanghai, but also a significant agenda for global cities. It is well known that the policy and theory concerning urban regeneration emerged under the pressure to improve living conditions and the urban environment in British and American cities in the 1950s. However, in the mid-19th century, Haussmann's renovation of Paris in the Second French Empire already displayed the classical feature of modern urban renewal. Facing social, political, and economic transition and renovation on the one hand and the pressure of population growth on the other, Haussmann's renovation of Paris included the adjustment of urban planning, improvement of living conditions, construction of infrastructure, landscape design, and real estate development led by the capital markets, finance industry, and consumer culture. The ideal, contradiction, and conflict that emerged during the process of such urban mutation can still inevitably and continually be found in today's urban renewal and regeneration. Since the 1960s, the rethinking of urban development, regeneration, and urban life has changed people's minds. The expectations for the city of the future are shifting toward justified, diversified, and sustainable development. There is a transformation taking place from the destroy–construct model to regeneration and renaissance, from land development to community building, and from economic benefit first to humanistic care about urban life.

SIBP New Place, named "Columbia Circle" by its developer, is a typical case of an urban regeneration project, with multiple historical, geographical, and cultural dimensions. Since its history began with the building of the Columbia Country Club 100 years ago in 1924, it has witnessed great changes and is worth studying for its architectural, urban, and humanistic value, and it is interesting to discuss its cultural significance as an urban space. The place was a significant part of the historical Columbia Circle in the 1920s and 1930s and regained the name "Columbia Circle" after its renewal in the 2010s because of the developer's wish to benefit from its historical and cultural associations, even though the historical Columbia Circle district is unrecognizable today. Three kinds of spatial models can be seen over the 100-year-long urban transformation of the place, each of which is related to a different concept of the ideal place. First, the model of the garden suburb was the original idea in the 1920s and 1930s. This gave way to the model of the *danwei* (work unit), as the working and living campus of the Shanghai Institute of Biological Products from 1951 to 2016; during this phase, the ideal was one of early collectivism. Finally, the community model came to be regarded as the basis for the transformation and creation of the SIBP New Place. These three models represent people's different visions of and actions with respect to the ideal living space during these different periods.

2. The Garden Suburb: 1920s–1930s

As one of the most outstanding garden suburb communities in the 1920s and 1930s, Columbia Circle was mainly planned and constructed on Shanghai's western periphery beyond the International Settlement and French Concession. The development of Columbia Circle, shaped by both the developer and the architect, reflected a transformation from the local Chinese countryside to one of modern Shanghai's emerging cosmopolitan garden suburb communities (Figure 1).

"Columbia Circle" specifically refers to the real estate project

Figure 1
Plan of Shanghai, 1928 survey by Shanghai Municipal Council. D=Western District of International Settlement. Columbia Circle (red dot on the map) was in the area of F. The areas H, G, and F shown in the map were proposed planned extension areas of the International Settlement.

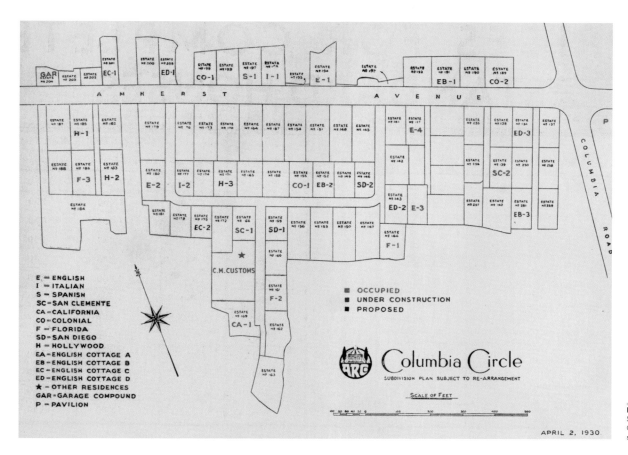

E = ENGLISH
I = ITALIAN
S = SPANISH
SC = SAN CLEMENTE
CA = CALIFORNIA
CO = COLONIAL
F = FLORIDA
SD = SAN DIEGO
H = HOLLYWOOD
EA = ENGLISH COTTAGE A
EB = ENGLISH COTTAGE B
EC = ENGLISH COTTAGE C
ED = ENGLISH COTTAGE D
★ = OTHER RESIDENCES
GAR = GARAGE COMPOUND
P = PAVILION

Figure 2
Subdivision plan of Columbia Circle, showing different styles, April 1930.

located at the west of the intersection of Amherst Avenue and Columbia Road, developed between 1928 and 1932 by Asia Realty Company (a major American developer in 1920s in Shanghai). The "Columbia" in the name was derived from the "Columbia Country Club" built in the 1920s on the north side of Columbia Circle.

The estate was on the west of the Western District, which belonged to Fahua District. It was planned as a potential extension area for the International Settlement in Shanghai. Although it was outside the International Settlement, this area had gradually become Shanghai's de facto "backyard" through the Municipal Council's illegal road construction, called the Grand Road Scheme, between 1924 and 1925. The Municipal Council built two main roads in this area in 1925, Amherst Avenue (now Xinhua Road) and Columbia Road (now Panyu Road). The construction of extra-settlement roads led by the Municipal Council of Shanghai's International Settlement began in 1862. The reasons behind the road construction were multiple, but among the most important motivations were the benefits of real estate development and the growing value of the land.[1]

The road construction conveniently connected the suburban area with the downtown. In the same year, Frank J. Raven, the chairman of Asia Realty Company, purchased the land. Raven was an American engineer who had come to Shanghai in 1904 to work for the Department of Public Works of the Shanghai Municipal Council and stayed on in the city as a real estate developer.

Raven did not develop the land immediately but waited for the real estate market to improve. In 1928, the Hungarian–Slovakian architect László Hudec[2] was appointed chief architect of the estate. The subdivision plan of Columbia Circle was largely indebted to America's garden suburb tradition. Despite the name, the overall layout was more a cluster of detached, similarly scaled residences in different

Figure 3
Life of 57 Columbia Road, the Hudecs' residence.

Figure 5
A bird's eye view of the Columbia Country Club in 1927.

Figure 4
This bird's eye rendering shows the whole Columbia Circle in its enlarged sense in 1930. The blue line defines the subdivision plan of the estate, and the yellow line defines the boundary of the Columbia Country Club. Edited by Feng Li.

styles than a formal "circle," reflecting the fashionable suburban lifestyle of the 1920s and 1930s (Figure 2).

The Columbia Circle project achieved great success. According to the Asia Realty Company annual report for 1931, a number of buyers, mainly upper-class foreign expatriates, purchased these residences in the spring and summer of 1931.

In 1930, apart from the Columbia Circle project, Hudec designed three-story, English-style houses for a development called the "Great Western Road Circle" for Chinese owners. In the same year, Hudec also designed a weekend bungalow for Frank Raven, as well as a residence for himself at 57 Columbia Road. The Hudecs' residence was restored in 2012, and the Hudec Memorial Hall has since been established on its ground floor. At the same time, Hudec also designed a more spacious house for Sun Fo (or Sun Ke)[3] near his own house (Figure 3).

The broader concept of Columbia Circle was more a lively community circle than a real estate project. Geographically, it referred roughly to the area including the Columbia Country Club, a riding school nearby, and the garden villa estate around it (Figure 4).

Designed by Shanghai's preeminent American architect, Elliott Hazzard, in 1923, the Columbia Country Club was a joyful place where Americans in Shanghai came to socialize with one another in the 1920s and 1930s. The club was, in a real sense, the cornerstone of the Columbia Circle community. As Patricia Luce Chapman, one of the earliest residents of Columbia Circle, recalled, "Most of our neighbors came over to my parents' parties, or bridge and poker games, and my parents to theirs. We loved the sports and the parties in the Columbia Country Club. Tennis courts, a swimming pool, and a squash court were there for us, and a large bowling alley. Under the Spanish–Mexican style arcade was a long verandah for dining and dancing."[4] (Figure 5)

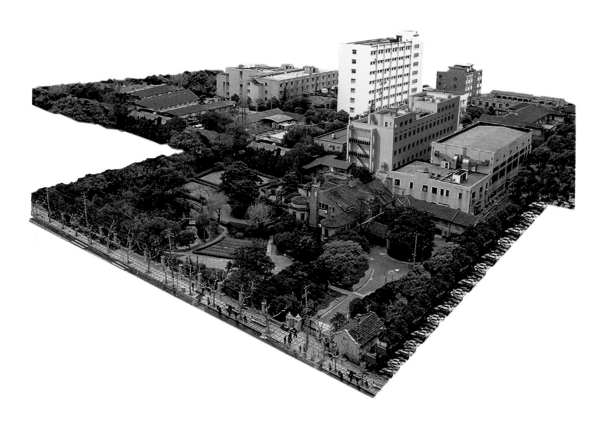

Figure 6
An overview of the SIBP campus.

In the 1930s, most of the residents of Columbia Circle were upper-class foreigners. By referring to the documents of the China Hong List, the foreign residents and their positions at Columbia Circle between 1930 and 1941 can be identified. Their jobs ranged from diplomats, dental surgeons, agents, attorneys, financial advisers, and engineers to journalists. The garden suburb community cluster could, therefore, be regarded as a foreign enclave in Shanghai's western suburbs. Among them, the most famous resident was J. G. Ballard, the well-known Shanghai-born English writer. Ballard once recalled in a private letter about his childhood, "…the French Club and the Country Club, where I suppose I spent the happiest days of my childhood, and to know that I could actually swim in the same pool."[5]

The idea of a country house built in a garden suburb to be enjoyed by upper-class people as a relaxed and comfortable escape from the noisy and crowded downtown has a very long history, reaching back even to the early Middle Ages in Europe. The world in a garden suburb is full of quietness, health, and freedom, like an attainable utopia in real life. When the Columbia Country Club was built in 1924, farmland on the south side of the Great Western Road (now the West Yan'an Road) stretched to the horizon. This rural landscape was obviously taken as a part of the dream life of Columbia Circle. For those foreign residents, the place may not just have been a quiet dwelling district far from downtown but also an imaginary homeland.

However, the ideal suburban place did not last for long. When the Pacific War broke out in 1941, the Japanese interned nationals of the Allied countries who were in China. The Columbia Country Club was used as a transit camp beginning in 1942 and became an internment camp in May 1943. The same club with a pool also saw its dark age in late 1941, when foreigners in Shanghai were interned by the Japanese. Many foreign residents who lived in the area of the Columbia Circle were put in the camp. As Ballard recalled, the Allied nationals in Shanghai "were bussed into Lunghua from our assembly point at the American Club (Columbia Country Club) near the Great Western Road, the large crowd of Brits, many of the women in fur coats, sitting with their suitcases around the swimming pool, as if waiting for the water to part and lead them to safety."[6] The dream place of enjoyment had become a prison. This cosmopolitan community in western Shanghai, therefore, collapsed during the war.

3. *Danwei*: 1951–2016

In 1945, after the Japanese army was evacuated from Shanghai, the Columbia Country Club building was taken by the Nanjing National Government of China and used as an office and lab for epidemic prevention. This is probably why soon after the foundation of the People's Republic of China, this place became the address of the Shanghai Institute of Biological Products. In a shift from a garden suburb dream place to an ideal collectivist campus, it became a *danwei*, which was originally a labor organization but soon integrated working and living spaces to become a combination of an institution and a home. The place underwent great changes from previous decades. These changes resulted in the first place from the urban development that was transforming the city of Shanghai as it embarked on the path to rapid urban modernization, but were also determined by a shift in the concept of the ideal place. If the natural landscape of the suburb garden was an ideal living space for those foreign residents, allowing them to recall the memory of their hometown or to imagine a dream life out of the crowded city, the *danwei* was a practice of an idealist society, of a future led by idealism. It was not only an embodiment of political,

Figure 8
The historical transformation
of the SIBP campus.

Figure 7
The main lab building of the SIBP.

economic, and social institutions but also a state-owned collectivist organization, an identity, and a benefit for employees. It implied a division of city and country and brought an advantage to laborers working and living in the city, who were given more benefits than farmers. *Danweis* were mainly set up in the city and offered a guarantee to citizens for their identity and life. *Danweis* were always self-owned, giving them an exclusive and autonomic place in the city. As a big institution, the SIBP had created a closed campus based on the site of the Columbia Country Club, and at the same time, created a spatial network on an urban scale.

The Shanghai Institute of Biological Products was founded in 1951 when it moved to the address of the Columbia Country Club from No. 222 Tiantongan Road. The institute was formerly known as the Shanghai Biological Products Factory, which belonged to the East China People's Pharmaceutical Company. After being given the name SIBP, the institute became a confidential research institution working on epidemic prevention for six provinces in East China and the city of Shanghai. In 1952, faced with biological weapons used by the US army during the Korean War, the Chinese government began to pay great attention to the development of the biological industry. In this context, the SIBP was joined by seven other biological factories and expanded to a campus that was in use until the SIBP moved out in 2016. The campus covered an area of more than 40,000 square meters and included not only the Columbia Country Club building but also Sun Fo House and other buildings (Figure 6).

SIBP continually transformed and renewed the campus from 1951. New industrial buildings were planned in a grid and designed following the functionalist principle. In this period, the place was transformed from a foreign residential and relaxing garden suburb district into a functionalist working and living space, a socialist *danwei*.

Figure 9
A map of the urban network of SIBP.
1) The Columbia Country Club;
2) Sun Fo House;
3) Swimming pool;
4) Clinical reagent production building;
5) Nursery;
6) Staff dining hall;
7) Head office (No. 1262 West Yan'an Road);
8) School of Biological Products (No. 1326 West Yan'an Road);
9) Dormitory (No. 1628 West Yan'an Road);
10) Farm of animals for experiments (No. 926 West Zhongshan Road, and No. 350 Anshun Road);
11) Dormitory on Anshun Road;
12) SIBP Hongqiao Branch.

In the early period, the construction work included not only new industrial facilities but also the transformation of historical buildings. First, the gym was transformed into a culture medium lab, with the installation of industrial machines and new dormers on the roof, and then extended on the north side in 1956. The Sun Fo House was occupied by the administration office; its plans were re-arranged to settle the admin departments. The dining room on the ground floor became a meeting room, and a storage room on the second floor was even used to save guns for military training. As a collectivist *danwei*, besides working spaces, the construction of the SIBP also included service facilities such as a nursery for employees' children. The old swimming pool of the Columbia Country Club was kept and continued to be in use, normally opening in summer for employees — it was the only space to keep its original function. In 1958, for reasons related to the Chinese political movement, the Great Leap Forward, the eight-floor main lab building, designed by Guo Bo, son of the famous Chinese scholar Guo Moruo, was built to show the power of science and technology (Figure 7). The site had previously been the club's football playground. This building was one of two buildings over five floors high that were constructed during that period in Shanghai. Over the more than sixty-year period of SIBP's operation, the place was transformed, with a mixture of historical and modern buildings and went from being an open, suburban landscape to a gridded, mixed-used campus (Figure 8).

After the 1980s, China stepped into a new period of renovation, not only in terms of political transformation but also of a cultural renaissance. The historical buildings in the SIBP campus were recognized for their historical and cultural value. As a diplomatic space for international visitors and international cooperation and exchange events was needed, the Columbia Country Club was refurbished to receive foreign guests. In the 1990s, the Club and the Sun Fo House were both included in the list of Shanghai's Excellent Historical Buildings. In 2003, the Sun Fo House was refurbished back to its original features and ceased to be used as an office building.[7]

For citizens living around the place, the SIBP campus was a secretive and mysterious place, a forbidden zone gated and separated from the city. However, for SIBP's employees, in their collective memory and mental cognitive maps, SIBP was not simply a closed block, but a spatial puzzle that included functional pieces which were distributed in the city like spatial fragments and grew as a part of history following the urban development of Shanghai (Figure 9). In the early period of the 1950s, Shanghai was planned to be an industrial city. This goal brought a dramatic and rapid urban transformation. Against this background, many street blocks and buildings in the district of Yan'an Road, Fanyu Road, and Xinhua Road were transferred to join the SIBP territory. For example, the building at No. 1628 West Yan'an Road, previously used as an administrative building by the Shanghai Labor Union, was transformed into a dormitory building for SIBP employees. It was later demolished in 1995 for the construction of the Yan'an Road expansion. To give another example, the former French Country Club on No. 1326 West Yan'an Road was used as SIBP's entertainment area and dormitory and then later transformed into the SIBP School in 1959. On the west side of the SIBP campus, there was originally a farm and some dormitory buildings at No. 926 West Zhongshan Road and No. 350 Anshun Road; later, these became the site for the SIBP Hongqiao Branch in the 1980s. The transformation of urban spaces in this area is part of the mutation of metropolitan Shanghai, and is also combined with the growth of the SIBP, with the transformation of the spatial organization model from that of the garden suburb to the *danwei*, as a product and representation of collectivism that can

33

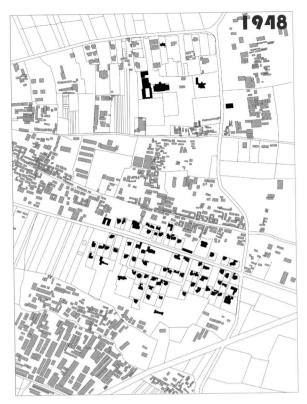

Figure 10
The transformation of the
urban context from 1948 to
2016.

create an autonomous sub-network within the city. Such an autonomous sub-network is not only a social institution and an organization of people but also a spatial production of social life.

In the latter half of the 20th century, the residential area of Columbia Circle was also involved in drastic urban transformation, and significant changes occurred, mainly in usage and ownership as well as the physical environment. After the founding of the People's Republic of China in 1949, according to the new Land Law, any property bought by foreigners on land outside the foreign settlements was illegal. These properties were confiscated and taken over by the housing sector of the Shanghai government. Most of the villas were re-assigned to local families. Meanwhile, many factories poured into this area, occupying land that was originally farmland around the Columbia Circle area. From the 1970s and 1980s, 17 six-story public residential units were built in succession, scattered throughout all the empty land in Columbia Circle. After 1990 and especially during recent years, a number of villas were transformed into private residences, company offices, restaurants, service apartments, and hotels.

The villas of Columbia Circle experienced wartime takeover during the 1940s, the process of socialization and subdivision in the 1950s, and a chaotic period during the Cultural Revolution. Fortunately, most of the villas still survive and have become part of Shanghai's listed urban vernacular heritage. However, the features of both the community and space of Columbia Circle as a whole in the periods of the 1920s and 1930s have already vanished into history (Figure 10).

4. Community: 2018 to Now

In the early 21st century, the place that had been home to the Columbia Country Club opened a new page in its history. The transformation of SIBP New Place is an urban re-development project produced by the Vanke Real Estate Company. The SIBP campus has been transformed into a creative park with high-level offices and commercial facilities such as restaurants, cafes, and shops. The project was completed in 2018. Its architectural concept design was proposed by OMA, and its detailed design was finished by local architects Arcplus Group PLC. The concept design for the landscape came from West 8 and was then completed by design firm LANDAU. In addition to being intended for business development, the transformation of SIBP New Place is also supposed to take into consideration the issue of community development. This is not only because the place is surrounded by residential blocks, meaning that the project has to consider its relationship with the neighborhood, but also because it is seen as part of the urban regeneration of Xinhua Road Subdistrict. For example, it was collected into the Urban Design Festival 2018, which was an urban and community event organized by the Xinhua Road Subdistrict government and social organizations. The event tried to create a cooperation between a series of renewed places and local people in the Xinhua Road area and then build an integrated network for the community in the whole subdistrict.

The community renewal movement has recently been regarded as a significant social and political project by the Shanghai government and also as an important aspect of the urban regeneration of the city of Shanghai. The announcement of the project, "Walking in Shanghai—Micro-renewal Projects of Community Space," by the Shanghai Urban Planning Administrative Bureau in 2016 can be seen as a sign of the importance of community renewal. The focus on the issue of the community can be understood as a result of Shanghai's urban development shifting from a phase of urban expansion to one of the urban renewal of existing environments on the one hand, and,

Figure 12
The SIBP New Place.

Figure 11
Bird's eye view of the
SIBP New Place.

on the other, it is a spatial practice for improving the management of grassroots governance. Xinhua Road Subdistrict, where the SIBP New Place is located, has a particular importance as a demonstration area for urban regeneration and community building in Shanghai.

The SIBP New Place is now an attractive public space for surrounding residents and Shanghai citizens. The mixture of historical buildings and modern architecture, the comfortable pedestrian environment, and various cultural events make the place a hot point in the city. The symbiotic relationships of different people and activities benefit from the project's spatial character of multiplicity. The place was re-planned, with different plazas and courtyards, following OMA's concept. Some industrial buildings were removed to leave space, but other modern buildings were saved and refurbished. Historical buildings were restored to their original features by a professional team. The mixture of historical, modern, and contemporary architecture makes the project a meaningful place. This design strategy brings the possibility and potential to allow various cultural contents to co-exist simultaneously in the SIBP New Place. For example, the entrance plaza facing the Spanish-style façade of the fashion club, which was previously the gym connected with the swimming pool, is a place for fashion and art events; the courtyard along the Columbia Country Club is now used by Tsutaya Books and is a relaxing place to have a coffee and rest; another courtyard behind it, covered by trees, is a quiet place drawing both older people and children to play. The SIBP New Place is open twenty-four hours a day to the public and is not a gated campus. The capacity for mixture and multiplicity of spaces in the SIBP New Place guarantees the creation of a positive and open public space for the Xinhua Road community and downtown Shanghai, as opposed to a limited luxury consumption area.

The opening of the SIBP New Place in 2018 means the beginning of a new history for this 100-year-old place. Through the transformation of existing buildings and redesign of the spatial environment, what is introduced is not only a change of function and increase in business value, a change from an institutional campus to a creative park of offices and commercial facilities but, more importantly, a transition of spatial identity and role, a transition from a secret and isolated street block to an open public space, which had been lacking in this area. The charm of contemporary urban life comes from the landscape of kaleidoscopic information but also the new community created by multiple social networks. Living with the sense of the fluidity and impermanence of a metropolitan city, people are continually connected and re-connected with the huge, complicated, and unrecognized city through the network nodes built by public spaces. Such connections help people recognize the city and also play the role of a new community in promoting social communication to take place. People meet each other in public spaces, join in events, and involve themselves in urban life. The memory of the past hundred years of history and spatial transitions support the SIBP New Place in being an urban node containing multiple connections with different people. In this case, the place is not only used by the surrounding residences as a shared, free open space for community building in Xinhua Road Subdistrict, but is also a fresh, active node in the social network of contemporary Shanghai (Figures 11, 12, 13, and 14).

5. Conclusion

The place has witnessed three great spatial and cultural transformations. The first was the construction of the Columbia Country Club and Columbia Circle, which involved a

Figure 13
The SIBP New Place.

Figure 14
The SIBP New Place.

transformation from the Shanghai countryside to an alien community living in a garden suburb. In its second period, it became an isolated but idealist courtyard and autonomous *danwei* of SIBP within but also out of the urban context. Finally, it has been transformed into a creative park and a public space, which is named the SIBP New Place and recalls the name of "Columbia Circle" with a vision of the contemporary community renewal of Shanghai. This story shows how architecture, urban space, and people's lives have changed over history. Furthermore, the history of Shanghai's Columbia Circle can help us understand the different concepts of the ideal place and ideal life hidden behind the dramatic transformation of the place, the city, and society.

Notes

1. Zhang Peng, Dushi xingtai de lishi genji — Shanghai gonggong zujie shizheng fazhan yu doushi bianqian yanjiu [Physical Foundations of City Form: A Study on the Relationship between Municipal Construction and Transformation of Urban Space in the International Settlement of Shanghai] (Shanghai: Tongji University Press, 2008), 144.

2. László Hudec was a Hungarian–Slovak architect who was active in Shanghai from 1918 to 1945 and was responsible for some most notable structures in the city.

3. Sun Fo (1891–1973) was a high-ranking official in the government of the Republic of China. He was the son of Sun Yat-sen, the founder of the Republic of China.

4. Patricia Luce Chapman, Tea on the Great Wall: An American Girl in War-Torn China (Hong Kong: Earnshaw Books, 2015), 139.

5. J. G. Ballard, "Looking back at Empire: The Background to Empire of the Sun," The Guardian, March 3, 2006. https://www.theguardian.com/books/2006/mar/04/fiction.film.

6. Ballard, "Looking back at Empire."

7. The information on SIBP's history comes from a series of interviews conducted in 2020 with Li Qianghua, Hu Jianping, Chang Guocai, Bao Xiaoli (retired employees of the Shanghai Institute of Biological Products), and Ma Lili and Hu Shensheng (senior residents living around the Shanghai Institute of Biological Products) by Hua Xiahong, Li Yingchun, Wu Jiao, Lin Xiaodan, Qin Yujie, and Mei Mengyue.

Figures

Figure 1: Plan of Shanghai, 1928 survey by Shanghai Municipal Council. D=Western District of International Settlement. Columbia Circle (red dot on the map) was in the area of F. The areas H, G, and F shown in the map were proposed planned extension areas of the International Settlement (Source: http://www.virtualshanghai.net/Map.php?ID=33).

Figure 2: Subdivision plan of Columbia Circle, showing different styles, April 1930 (Source: Columbia Circle, Promotional Brochure, Hudec Collection, Library of University of Victoria, Canada).

Figure 3: Life of 57 Columbia Road, the Hudecs' residence (Source: Hudec Collection, Library of University of Victoria, Canada).

Figure 4: This bird's eye rendering shows the whole Columbia Circle in its enlarged sense in 1930. The blue line defines the subdivision plan of the estate, and the yellow line defines the boundary of the Columbia Country Club. Edited by Feng Li (Source: Columbia Circle, Promotional Brochure, Hudec Collection).

Figure 5: A bird's eye view of the Columbia Country Club in 1927 (Source: hpcbristol.net).

Figure 6: An overview of the SIBP campus (Source: The Shanghai Institute of Biological Products, photo by Hu Jianping).

Figure 7: The main lab building of SIBP (Source: The Shanghai Institute of Biological Products).

Figure 8: The historical transformation of the SIBP campus (Source: drawing by Qin Yujie and the Historical Building Preservation Design Institute of Arcplus Group PLC).

Figure 9: A map of the urban network of SIBP (source: drawing by Mei Mengyue, Wu Jiao, and Hua Xiahong).

Figure 10: The transformation of the urban context from 1948 to 2016 (Source: drawing by Lin Xiaodan, Hua Xiahong, Feng Li, Wu Jiao).

Figure 11: Bird's eye view of SIBP New Place (Source: photo by Historical Building Preservation Design Institute of Arcplus Group PLC).

Figure 12: SIBP New Place (Source: photo by Historical Building Preservation Design Institute of Arcplus Group PLC).

Figure 13: SIBP New Place (Source: photo by Historical Building Preservation Design Institute of Arcplus Group PLC).

Figure 14: SIBP New Place (Source: photo by Historical Building Preservation Design Institute of Arcplus Group PLC).

Learning in Doing:
The Role of the Architecture
School in Urban Regeneration

YAO Dong, Tongji University, China

Author's Information
YAO Dong: yaodong@tongji.edu.cn

Abstract

Rapid urbanization and an aging population have changed Chinese cities tremendously over the past four decades. In line with other international pilot programs on participatory design, Tongji University established the first service-learning studio on the Chinese mainland in 2017. The project site, Nandan Subdistrict, is a super dense and aging neighborhood in Shanghai's city center. The service-learning team became involved in community planning and video recorded a whole year's transformation. As a result, the new center is being used efficiently by neighbors of various ages and has turned into the most popular place in the neighborhood. After a whole year's involvement in the neighborhood, there are three main conclusions: many former participants turned into supportive partners; the service-learning team of faculty and students had a significant advantage in community planning; community planning should design not only the space but also a creative way of place-making.

Keywords

Community planning, service-learning, participation, architecture school.

1. Introduction

In the context of China's rapid urban development and aging population, it is not surprising that behind the shiny high-rises, there are often run-down spaces and groups of older adult citizens who seem to have been left behind by the fast-changing city. These areas are like the city's calluses, waiting to be discovered and softened. However, there is often a disconnect between the architects who participate in urban regeneration and the residents, property owners, and local governments of such communities. To solve this problem and promote the participation of the School of Architecture in urban renewal, Tongji University set up a service-learning studio to explore new ways in which teachers and students can participate in urban renewal, promote public participatory design, and enable students to "learn by doing." This paper records the studio members' participation in community renovation throughout the year in the hope that sharing their experiences with the service-learning course will help its lessons to be applied more widely.

2. Urban Decay and the Aging Society

Four decades of rapid development have changed China. Most cities have undergone tremendous changes. Urban regeneration, the ubiquity of automobiles, and high-intensity development have often been in conflict with the interests of the aging population. These three phenomena have accelerated the shrinking of the social life of the older adult and together represent an unavoidable new problem.

2.1 Urban Regeneration
Urban regeneration has resulted in many older adult people becoming "strangers" in the city. The renewal of the city often results in the disappearance of its existing spatial relationships, and many critical urban landmarks become absorbed into a nameless and unrecognizable spatial environment. At the same time, urban regeneration also destroys the spatial memory and life experience of the older adult.

2.2 The Ubiquitous Automobile
The ubiquity of the automobile has turned many older adult people into "disabled people" living in the cities. Having become accustomed to the pedestrian-scale construction historically used in districts designed for pedestrians, non-motorized vehicles, and bus transportation, they are now confronted with newly built super-blocks and wide roads built for the benefit of businesses and private transportation. This change of scale is not a great problem for young people, who can adapt to new transportation types such as private cars and subways. Still, it is undoubtedly an extremely large problem for the older adult, who mainly use traditional methods of transportation such as walking.

2.3 High-Density Development
High-density development is turning many older adult people into "outsiders" in urban life. With the adoption of the high-density development model, the per capita green space and per capita development space in urban centers have been continuously reduced. Open outdoor public spaces have been replaced by large shopping malls, high-end residential areas, and other "public spaces in private areas." Newly built green spaces are often located on the outskirts of cities, which are difficult for ordinary older adult people to reach. In contrast, gorgeous new spaces in cities are often accompanied by a consumption orientation or access control facilities that discourage ordinary older adult people from using them. The overcrowding of older adult people in parks in many cities and the conflicts over square dancing with community residents are all caused by this. With nowhere to socialize, more and more older adult people are excluded from urban life.[1]

2.4 Public Service Facilities Mismatch
In the 100 years since its opening to international trade in 1843, Shanghai has grown from a southeastern metropolis of 520,000 people to China's largest city.[2] As the population and the economy grew, urban renewal aimed at public services, transportation infrastructure, and urban functions began in the late nineteenth century. The construction of public service facilities effectively alleviated the expectations for public space for the whole of society, but it also exposed a series of mismatch problems. Some facilities that are considered very necessary by architecture schools are inefficiently used in real life. In contrast, community residents often welcome other facilities that are considered unhealthy or likely to cause conflict.[3]

3. Participatory Planning

3.1 Multiple Cooperation and Community Restoration
The reconstruction of old cities before the industrial revolution often highlighted the legitimacy of the rulers through commemorative means. With industrial development, the urban population began to accumulate rapidly, which brought about urban problems such as the deterioration of public health, traffic conditions, and the living environment. Under the influence of environmental determinism, the top-down reconstruction of old cities was used to promote the modernization of the urban environment. In the middle of the 20th century, to solve the problems of inner-city recession, deterioration of living conditions, traffic congestion, and poverty, Western developed countries began to adopt the clearance of slums in old cities as the main means of urban renewal. Since the 1960s, criticism of and reflection on top-down urban renewal have resulted in a new approach to old city reconstruction, involving cooperation between multiple stakeholders and community restoration. The built heritage, public participation, transportation, and public space have been gradually incorporated into the areas of cooperation in the reconstruction of old cities.

3.2 Participatory Planning in Architecture Schools
Architecture schools have used participatory planning as a new way to intervene in urban renewal since the 1960s. First, the Pratt Institute established a community design center in New York City to serve minority residents. In the United Kingdom, Rod Hackney, a doctoral student and architect at the University of Manchester, united the neighborhood community at Black Road in Macclesfield and succeeded in promoting redevelopment from the bottom up. They encouraged public participation worldwide, and community planning gradually became popular in transforming old cities.[4] To reduce accidents and restore the street's public space function, Professor Nick de Boer of the Delft University of Technology proposed the concept of a living courtyard (Woonerf), which is a garden transformation of roads in residential areas that has been replicated around the world since 1969.[5] In the transformation of the core district of Manhattan, Janet Sadi-Khan, New York's transportation commissioner, promoted an experiment in "Streetfight" with public

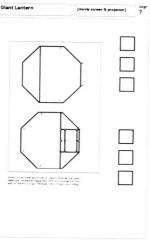

Figure 1
Night Market Studio,
a service-learning course in
Washington University,
Seattle.

participation. She made Times Square Broadway a non-motorized area, built more than 60 city squares, and created a world-class public space for New York.[6]

In architecture schools in the United States and worldwide, there has been vigorous development of the community planning/design studio approach in combination with the philosophies of "learning in doing" and "service-learning" in the context of various majors, including landscape architecture and urban planning. (Figure 1)

4. Service-Learning

4.1 The Deterioration of Autonomous Learning Ability

The deterioration of autonomous learning ability has become a common problem among architecture undergraduates in recent years. The resulting lack of ability manifests itself in various professional problems. Although the student can, in general, grasp the knowledge imparted in the classroom, their progress in dealing with the types of problems that need to be approached through independent study tends to be stalled. On the surface, senior students are mostly able to master the concepts behind design methods, conduct case studies, and learn to use the proliferation of drawing software tools and technical applications. However, there has been a constant regression in the depth of actual learning: the phenomenon of "attaching importance to homework performance while neglecting process deepening" is becoming more common, and the discovery of design problems, the accumulation of processes, and the optimization of technical links are generally missing.

The degradation of autonomous learning has a knock-on effect. Superficially, the work of senior undergraduate students stagnates at the level of junior students. However, at a deep level, there is a lack of

ability with respect to a series of basic issues in architecture. The site environment is replaced by red lines representing the site in the plans. Equal communication with partners and users is replaced by the task book. Renderings and concepts replace creative solutions to specific problems. As classroom teaching fails to promote independent learning, it has become a kind of paperwork disconnected from life. Inadequacies with respect to areas such as environmental awareness, communication skills, and individualized problem-solving abilities greatly hinder architecture students' future growth.

4.2 The Origin and Significance of Service-Learning

Service-learning can not only help students establish a sense of social responsibility, but is also a beneficial teaching method in terms of enhancing independent learning ability. Service-learning refers to a teaching method that realizes experiential learning through community service. It is "a form of experiential learning with service as the carrier. Service-learning applies to all stages of learning from primary schools to universities, enhancing the effectiveness of learning, engaging students actively in experiential learning in curriculum-related contexts,"[7] and forming a lifelong bond between students and their community.[8] This term, first proposed by the Southern Regional Educational Board in 1967,[9] contains the two core contents of "community service" and "experiential learning." The historical tradition of American higher education's involvement in community service can be traced back to the Land Grant Act of the 1860s. The experiential learning theory was put forward by the educator John Dewey: "Students not only learn a lot of knowledge and skills that cannot be provided in the curriculum in experiential activities but also experiential activities provide students with the practical application of the knowledge

39

learned in the classroom. An opportunity to organically link the knowledge of various disciplines."[10] Unlike pure community volunteer service, service-learning emphasizes the ability to apply knowledge through reflective learning in the process. "Learning in doing" can help students gain a sense of social responsibility, and integrating reality into their understanding of things can enhance their independent learning ability. As the old saying goes, "Give a man a fish and you feed him for a day; teach a man to fish and you feed him for a lifetime." The aims of setting up service-learning courses in architectural learning are to shift from case-based architectural learning to autonomous discovery of problems and to apply the knowledge and skills learned in the past to solve them.

4.3 Service-Learning Courses in Architecture-Related Majors
Since the late 1990s, service-learning courses have been gradually introduced into architecture-related majors, starting in the United States and gradually expanding to all parts of Asia. For example, the Department of Landscape Architecture at the University of Washington (Seattle) opened the "Chinatown Night Market" course in 2006. In cooperation with the Chinatown youth organization WILD (Wilderness Inner-City Leadership Development), professional college students and non-professional middle school students jointly designed six groups of night market landscape installations for Qingxi Park. The program helped revitalize the Chinatown community, and both the professional and non-professional students expanded their knowledge and communication skills through collaboration.[11] The School of Design at Zhongyuan University in Taiwan launched a course on "Community Construction and Public Participation" in 2012. It cooperated with the Xiaoli Primary School in Meinong District and Tao Jianbao Community in Taoyuan County to investigate and

draw a large community resource map. The Real Estate Department of Hong Kong Polytechnic University launched a "Community Housing" course in 2015, which investigated the actual housing problems of different income groups in the Aimin Village Community of He Wentian and put forward research suggestions for future housing policies. After receiving skills training, the students not only completed a reflection report, but also completed simple housing repair services in low-income older adult families. The experience of these courses can serve as a useful reference while implementing service-learning courses as part of the architecture major.

5. Design for All: The Project at Nandan

After a year of preparation, a service-learning studio was established at Tongji University in the spring semester of 2017. As an elective course open to Grade 4 students in the Bachelor of Architecture program, the studio emphasized two major pursuits: community service and independent learning.

5.1 Nandan Subdistrict
The students enrolled were required to do both design and community service in a dilapidated neighborhood. Located in Shanghai's city center, Nandan Subdistrict, the selected neighborhood, is characterized by its super-high population density and aging ratio. (Figure 2) After many rounds of old city reconstruction since the founding of the People's Republic of China, from *gundilong* (simple thatched shacks built by the poor people of Shanghai before liberation) to shanty houses, and then to the residential community pattern formed in the 1990s, Nandan Subdistrict has witnessed the historical change of Xujiahui from an urban–rural junction to a sub-

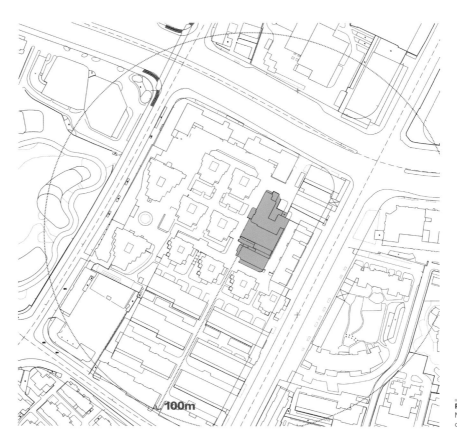

100m

Figure 2
Nandan Subdistrict and the community center site.

Figure 3
The rendering for the Exchange Day proposal.

center of Shanghai. The area was originally located on the periphery of Xujiahui and was home to socially marginalized workers and unemployed people living in the *gundilong* shantytown. In 1950, it underwent its first reconstruction, which involved paving roads, laying pipes, dividing land, and building earthen houses. After that, residents gradually rebuilt two-to-three-story simple houses according to their economic conditions. In 1992, it was incorporated into the "365 Project" and underwent a second renovation. The original houses were demolished and built into a complete set of six-to-seven-floor buildings. There are 19 six-story housing slabs and four high-rise housing towers. Gathered in a 3.5 sq. hectare site, most of the units are below current standards. Some units are smaller than 30 sq. meters, with up to six residents. The population density is about 170,000 per sq. kilometer. There are 2,012 residents aged 60 and over, accounting for more than one-third of the total population and demonstrating the emerging need for services for the older adult.

There is a primary school in the southwest corner of the plot. There is a one-floor public service facility building in the middle of the plot and one three-floors kindergarten building. On the north side of the road, buildings for public institutions were built in the open space obtained by the demolition, including four 18–27-story high-rise buildings for the public security system in the northwest corner, two multistory buildings for the fruit company in the northeast corner, and large multistory houses built by the postal and telecommunications system for retired cadres.

The levels of occupation, income, social status of residents, and living standards in the welfare housing unit are far greater than those in the resettlement housing, causing social differentiation in the same community plot. When it was included in the "Beautiful Homes Action Plan" in 2016, Nandan Subdistrict had already presented a typical

case of the "three highs and two lows" and of a service mismatch, with prominent social differentiation and neighborhood conflicts. The "three highs and two lows" refers to high housing density, high population density, high average age, low-income levels, and low per capita housing area. There are common environmental problems in the existing residential areas in the Nandan Subdistrict, such as environmental decay, the conflict between parking and greening, security risks caused by road occupation and parking, low efficiency of public facilities, and insufficient public space. There is a shortage of per capita living space and public space in Nandan Subdistrict. Public facilities and public spaces are also poorly used.

5.2 Course Arrangements for Service-Learning

To optimize the neighborhood's quality of life, a community center in the geographic center of the subdistrict was proposed. The assignment was composed of two steps. The first step was to organize an Open Day for the neighborhood to encourage participation and collect the neighbors' feedback. The second step was to propose a community center design with the necessary functions, as established from the Open Day events. The assignment originated from the studio instructor's real practice in designing a remodeling of a current facility and introducing new functions. Although the studio design proposals were fictional, the Open Day and questionnaire served as an organic component of the participatory design process. (Figure 3)

To encourage the development of autonomous learning ability, there were prominent differences between the teaching in this course and general core design courses in three main aspects, i.e., course pressure, task book, and grading method. The course was arranged in the self-selected stage of the senior year of

Figure 4
The Exchange Day, Nandan
New Village, 29 April 2017.

the five-year architecture undergraduate course, which greatly reduced the design course's study pressure. Compared with other core design courses, the teaching in this course involved a long cycle of questions over one semester. Moreover, students no longer had the pressure of other simultaneous courses, which fully guaranteed that they would be able to invest the time needed in learning. To avoid the limitations on autonomous learning imposed by a detailed task book, students were able to choose to design the whole project or part of its functions under the premise of a land area of 1,000 sq. meters, a building area of 3,000 sq. meters, and the function of community public service facilities and to reach the required depth of achievement and standard, according to their learning objectives. To avoid emphasizing the grading method based on drawing expression, the grading for the course was based mainly on the research and expression of the design process in the reflection report, and the design results were only used as an auxiliary.

The four-week preparatory phase focused on teaching community service skills and strengthening the principle of self-directed learning. The knowledge impartation included the introduction of the design and implementation scheme of the public service facilities in the Nandan Subdistrict, as well as lectures on design, community architecture, and participatory design methods, starting with the basics, with the purpose of emphasizing the logic of design thinking and operation, as well as the two-way design of space and behavior. The course assignments included portfolio reporting, and subsequent assignments included a base survey, case study, site design, and other topics. The purpose of the portfolio was to urge students to find their characteristics and shortcomings through writing summaries and reports and to set their personal learning goals. The follow-up homework was intended to help students

understand the principle that design work builds gradually and focuses on accumulation. At the same time, it was hoped that reviewing the topics in a real environment could help the students to reflect on their learning goals and on how to improve their abilities.

After the preparatory phase, the six-week service phase was the most intensive part of the course, consisting of three weeks of community interaction days to design the participation tools, one week of library quiz design, two weeks of preparation for the interaction day, and the last two days of interaction day activities. The participation tools included constructing space devices and organizing exhibition contents, posters, models, questionnaires, and other props. The students were divided into two groups, carried out three design scheme rounds, and finally integrated them into the implementation scheme. According to the personal learning goals they had set in the early stage, the students undertook specific tasks dealing with video recording, posters and questionnaires, interactive exhibitions, interactive models, exhibition boards, exhibition stands, or rest chairs. All kinds of materials were delivered to the site, and the foundation was set up the day before the community interaction day. The construction started early on the morning of the activity, and the collection of the demand questionnaire was completed. The above work involved constructing the site environment and communicating with residents, suppliers, and partners, which provided students with a rare opportunity to exercise their skills.

The last five weeks of the course were a reflective period. The community service work included sorting out the results of the survey on residents' demands, soliciting opinions, voting for micro renewal, running a model workshop, and gathering feedback opinions to support the optimization of the design for the

Figure 5
Public hearing, Nandan New Village, 16 July 2017.

Figure 6
The Neighborhood Festival, Nandan New Village, October 2018.

community public facilities. The assignment required the completion of the architectural design and of a reflection report on the community public service facilities. Implementing the principle of independent learning, each student was able to choose the design's scale and depth according to their interests and ability. However, all the designs needed to reflect on their learning over the past four years and the learning orientation of the course and to solidify this content in the form of a report.

6. The Partnership

6.1 Design Stage

Obtaining information regarding the residents' actual needs for public services, mobilizing community participation, and resolving conflicts were the main goals of the large-scale public participation activities in the design stage. On Saturday, April 29, 2017, an interaction day was held at the community center's crossroads, including projects such as a display of updated plans, case exhibitions of community-building achievements, model workshops, interactive games, opinion polling, and questionnaires. (Figure 4) The eight-hour activity spontaneously attracted about 300 community residents to participate, and 100 valid questionnaires were collected. About 10% of the respondents were 18 years old or younger, 22% were 19–44 years old, and 32% were 44–64 years old; more than 35% of the respondents had a bachelor's degree or above, and people living in groups of more than three people outside the context of the nuclear family accounted for more than 40% of respondents, meaning that the sample was fully reflective of the characteristics of the full range of ages eligible to participate.

6.2 Construction Stage

During the construction phase, the goal was to optimize design, prevent risks, and promote community integration. The holding of opinion consultation days and family gathering days enabled open communication about multiple subjects. On the morning of July 16, 2017, the second large-scale community participation event, an Opinion Solicitation Day, was held in the Neighborhood Club. (Figure 5) The participants in this activity included leaders, service office staff and other government representatives, a team of planners, grassroots representatives from libraries, neighborhood committees, community police, and 55 community residents of different ages. The street leaders introduced the detailed design of the Nandan Neighborhood to the residents, which subsequently triggered discussions on focal issues such as dance halls, spaces for playing chess and cards, and older adult care services. On January 18, 2018, the third large-scale public participation event, the Family Gathering Day, was held. This event included the trial operation of a series of functions such as community older adult care services, community study rooms, Chinese medicine rehabilitation, micro-theaters, community classrooms, smile dispaly walls, displaying posters, an exhibition on the history of Xujiahui, and other community square test content.

6.3 Operation Stage

The public participation during the operation stage served as a demonstration of the creative use of public space. The more important purpose was to build a community with a sense of subjectivity and sustainable development through neighborhood gatherings. The community planning team organized two neighborhood festival activities in the community square in March and October 2018, attracting more than 1,000 residents to participate. (Figure 6) The community square was divided into four areas by artificial turf: a stage, a market, a planting area, and a game area. Stage performances, movie screenings, youth club productions, and a potluck party were organized for residents of all ages, including young people and teenagers. There was also a second-hand market, nature education, and interactive games. The two neighborhood festival activities were the result of the cooperation of universities, neighborhoods, neighborhood committees, libraries, community nursing homes, and third-party social organizations. During the neighborhood festivals, the community public service facilities served as classrooms for civic education.

The organization of the community participation helped to mobilize all social resources' common participation in the renewal process and created the possibility for the residents to build a better living community. The public space invested in by the street is not only a communication platform for environmental remediation but also a stage for multi-governance, an incubation platform for community organizations, and a bond among community residents, encouraging them to help each other and pursue a better life together. The community pension service enterprise on the third floor of the neighborhood community undertook the entire facility's daily property operations. Neighborhood committees, libraries, social organizations, and autonomous groups provided a wealth of activities. The Nandan Subdistrict neighborhood was transformed into a public living room on the residents' doorstep throughout the year. It became a place for residents of different ages and interests to mingle. Opera lovers formed a "Huai Opera Troupe," dance and fitness experts formed a square dance team, and residents who like gardening formed a "Bottle-Garden" community. Together, they added vitality to the public space with their actions. In 2019, the implementation of "household waste classification collection" in Shanghai offered a good opportunity to test the community-building achievements. At the beginning of the garbage sorting, 87 residents volunteered as volunteers. A Nandan resident said, "The number of volunteers and their enthusiasm is the envy of the neighborhood next door."[12] Garbage classification has changed the traditional consciousness of the neighborhood. In the early stage of garbage classification, the residents were touched to learn that their implementation of the project in their multistory community was better than that of the high-rise community. Resident volunteers from high-rises began to act as self-initiated propagandists, going door to door to promote the policy. The volunteers came up with ideas to solve the problem that some office workers raised about the conflict between regular and fixed delivery times and their after-work hours.[13] Community renewal brought the Nandan residents together in pursuit of a better life. Residents who used to complain about problems began to learn to communicate and solve them first, as a Nandan resident said, "After all, everyone is for the whole community's good."[14]

7. Conclusion

The service-learning studio has transformed the architecture students. This type of teaching strengthens the relationship between the university and the community and enhances students' ability to learn independently in their professional fields. The latter is embodied in three aspects: base cognition, communication with people, and individualized problem-solving. Learning in doing,

the studio, and following the new facility's transformation turned out to a priceless class for both the students and faculty members. Participatory design has brought real demands to the students vividly, and all the participants have become their teachers. The students also contributed a great deal to the neighborhood. Without their participation, there would not have been so many events bridging the neighbors and designers, there would not have been a platform for discussion and negotiation, and it would not have been possible to meet diverse needs. This participatory design course of "learning in doing" provided a new way for the faculty and students of the School of Architecture to participate in community and urban regeneration. In summary, the studio changed the students and the neighborhood and transformed the methods of architecture pedagogy.

Notes

1. Yao Dong, "Mianxiang laolinghua de chengshi sheji—'Rouxing chengshi' de zaichanshu" [Soft City: Urban Design for the Aging Society], Urbanism and Architecture, (Mar. 2014): 48-51.
2. Cao Hongtao and Chu Chuanheng, Dangdai Zhongguo de chengshi jianshe [Urban construction in contemporary China] (Beijing: China Social Sciences Press, 1990).
3. Yao Dong, "Dachengshi 'yuanju anlao' de kongjian cuoshi yanjiu" [Spatial Countermeasures for Ageing-in-Place in Big Cities], Urban Planning Forum, (Apr. 2015): 83-90.
4. C. Richard Hatch, The Scope of Social Architecture (New York: Van Nostrand Reinhold, 1984); Wates Nick and Knevitt Charles, Community Architecture: How People are Creating Their Own Environment (Routledge, 1987).
5. Michael Southworth, "Street for Human Too," (Nov. 2019). http://www.architectureweek.com/2004/0505/building_1-2.html; Eran Ben-Joseph, "Changing the Residential Street Scene: Adapting the Share Street (woonerf) Concept to the Suburban Environment," APA Journal, Autumn (1995): 504-515.
6. Janette Sadik-Khan and Seth Solomonow, Qiangjie: Dachengshi de chongsheng zhi lu [Streetfight: Handbook for an Urban Revolution], translated by Song Ping and Xu Ke, (Beijing: Publishing House of Electronics Industry, 2018).
7. Zhou Jiaxian, "Meiguo fuwu xuexi lilun gaishu" [The Summary of Service Learning in America], Studies in Foreign Education, (Apr. 2004): 14-18.
8. Zhou, "Meiguo fuwu xuexi lilun gaishu," 14-18.
9. Tom Angotti, Cheryl Doble, and Paula Horrigan (eds.), Service Learning in Design and Planning: Educating at the Boundaries (New Village Press, 2011).
10. Zhang Tianjie and Li Ze, "Guanhuai tazhe / kuayue bianjie: Meiguo gaodeng yuanxiao fengjing yuanlin fuwu xuexi kecheng zouyi" [Seeing the Other, Crossing the Boundaries: An Introduction to Service-Learning Courses in Landscape Architecture in the American Universities], Chinese Landscape Architecture, (May. 2015): 27-32.
11. Zhang and Li, "Guanhuai tazhe / kuayue bianjie," 27-32; Zhao Xibin and Zou Hong, "Meiguo fuwu xuexi shijian ji yanjiu zongshu" [A review of service learning practice and research in the United States], International and Comparative Education, (Aug. 2001): 35-39.
12. Gong Danyun, "Laji fenlei, dui Shanghai de gengshenceng yiyi xiangguo ma?" [What is the deeper significance of garbage classification to Shanghai?] Shanghai Observer, July 21, 2019. https://www.jfdaily.com/news/detail?id=164547#areply.
13. Gong, "Laji fenlei, dui Shanghai de gengshenceng yiyi xiangguo ma?"
14. Gong, "Laji fenlei, dui Shanghai de gengshenceng yiyi xiangguo ma?"

Figures

Figure 1: Night Market Studio, a service-learning course in Washington University, Seattle, Credits: http://courses.washington.edu/nightmkt/.
Figure 2: Nandan Subdistrict and the community center site (author's photo).
Figure 3: The rendering for the Exchange Day proposal (author's photo).
Figure 4: The Exchange Day, Nandan New Village, 29 April 2017 (author's photo).
Figure 5: Public hearing, Nandan New Village, 16 July 2017 (author's photo).
Figure 6: The Neighborhood Festival, Nandan New Village, October 2018 (author's photo).

Skyrise Greenery as an Ecological Renewal Approach in Chinese Cities

DONG Nannan, Tongji University, China
ZHAO Shuangrui, Tongji University, China

Authors' Information
DONG Nannan: dongnannan@tongji.edu.cn, ORCiD: 0000-0002-7935-9619;
ZHAO Shuangrui: 1930115@tongji.edu.cn, ORCiD: 0000-0001-7788-7349.

Abstract

High-density cities in China face problems with their thermal environment, insufficient activity space, and fragmentation of their ecological landscape. Vertical greening technology has become a common approach in eco-city planning and green building design to meet the ecological goals of urban renewal. This paper first introduces recent developments in vertical greening in Chinese cities from the perspective of the policy development process. Then, combined with the experiments of the Innovative Urban Green (IUG) research group, the latest skyrise greenery projects in the urban renewal of Shanghai are summarized from the integrated aspects of planning tools, sustainable design, efficiency evaluation, economic benefits, and smart maintenance technology.

Keywords

Ecological restoration, vertical greening, efficiency evaluation, high-density city, urban renewal.

1. Background

China's rapid urbanization in the past 30 years has changed the relationship between cities and nature. According to statistics, from 1980 to 2010, China's urban–rural population gap gradually decreased, while the urban population continued to increase. Since 2010, the population living in cities has exceeded the population living in rural areas. The number of high-rise buildings in cities has increased in recent years, and the overall trend of urban development is towards concentrated vertical expansion. These high-density urban developments face environmental degradation problems, including insufficient ground-level greenery, improper management and use of urban rainwater resources, air pollution, and urban heat island effects. In addition, urban green space has been eroded, biodiversity has decreased, and habitat fragmentation has increased.

Therefore, in the process of urban renewal and restoration, special attention needs to be paid to increasing the quantity of green space. Vertical greening is especially suitable for high-density urban environments and is regarded as an important means of urban renewal.

1.1 Ecological Environment Problems during Dynamic Urban Development

Biodiversity is the manifestation and existence of biological factors in the environment. Urban biodiversity is mainly reflected in habitat and species diversity. In the high-density urban settlement environments, the loss of biodiversity caused by human activities is gradually intensifying with urban construction and expansion. This may lead to irreversible negative impacts on the environment, such as the fragmentation of biological habitats and the invasion of alien species. From the perspective of species invasion, it is necessary to use a landscape ecology approach to protect the biodiversity of the urban built environment and increase consideration of ecological planning and ecological design.

The expansion of the city disrupts the original balanced relationship between nature and human settlement, destroys the habitat at the city boundary, and at the same time, increases the importance of maintaining elements of the natural environment in the city center. Owing to the fragmented green space network within the city, vegetated landscape patches are often small and scattered. Ecological techniques such as three-dimensional greening can improve the connectivity of the landscape patches in the urban built environment and optimize the urban landscape structure.

Vertical urban greening has other ecological restoration benefits. For example, surface runoff in cities often transports pollutants to waterways and wetlands. Rooftop vegetation can reduce the total site runoff, intercept rainwater, and increase water penetration into the organic surface. In the central business district or downtown, active ecological roofs can eliminate the site runoff from small rainstorm events and reduce the pressure on integrated sewerage systems.

The formation of the urban heat island effect is also related to changes in the nature of the underlying surface of the city and the cooling influence of natural vegetation.[1] Urban buildings are often tall and dense with relatively small areas of urban green space and water bodies in the suburbs, and the hard environment reflects and absorbs heat multiple times. Over time, natural areas of soil, water, and land cover on the underlying surface gradually decrease, while impervious surfaces increase, leading to problems such as reduced surface water transpiration and accelerated runoff. In addition, greenhouse gases cannot be effectively absorbed in large amounts, which makes cities a heat source. Improving the vegetation coverage of urban surfaces is currently the most important response to the heat island effect in cities globally. Ecological measures include increasing trees and vegetation, green roofs, low-temperature roofs, and cool roads.[2]

Global environmental pollution arising from urban construction has become increasingly serious and endangers people's health. In recent years, statistics show that the incidence of respiratory diseases has been high in Chinese urban areas. The most frequently discussed pollutant concentrations in the literature are ozone, sulfur dioxide, particulate matter, and nitrogen oxide. Increasing urban greening can enhance the adsorption of dust on the surface of plant leaves, intercept particulate matter, reduce the concentration of pollutants in the air, and prevent additional air pollution. These changes can improve human health and create economic benefits by reducing human mortality.

1.2 Solving Problems through Ecological Urban Renewal

To solve the various environmental degradation problems faced by high-density urban areas, it is necessary to set up appropriate planning guidelines for ecological cities. Since 1992, China has proposed the strategy of garden cities. A national garden city has a balanced distribution, reasonable structure, perfect functions, beautiful landscape, comfortable living environment, and a safe and pleasant environment according to the standards of the Ministry of Housing and Urban-Rural Development.[3] Since the establishment of national garden cities in 1992, a total of 17 batches of 310 cities, 8 batches of 212 county towns, and 5 batches of 47 towns have been awarded the title of national garden city, county town, and town by the Ministry of Housing and Urban-Rural Development. More than 20 years of recognition of incremental urban ecological renewal has effectively promoted the development of urban ecological construction and improved the quality of urban livability.

In recent years, the ecological renewal of cities has evolved from the incremental development of garden cities to focus on the renewal of the high-density inner-city environment. Vertical greening, as an important and effective means of inventory renewal and ecological restoration, plays an important role in this process.

2 Development of Vertical Greening in China

2.1 Vertical Greening Development among Chinese Cities

China has published many policies and regulations on vertical greening. At the national level, these include: "Code for Acceptance of Construction Quality of Roof GB50207-2012," "Technical Code for Roof Engineering GB50345-2012," and "Technical Specification for Green Roof JGJ 155-2013."

Regulations have also been proposed for specific cities and provinces in recent years in China. Northern provinces like Hebei, Henan, Shandong, and cities like Beijing, Tianjin and Qingdao have proposed the following regulations: "(Beijing) Specification for Roof Greening DB11/T 281-2015," "Technical Regulations for Roof Greening in Hebei Province DB13/T-1433-2011," "Henan Province Three-dimensional Greening Technical Specifications DB41/T 921-2014," "(Shandong Province) The Technical Specifications of Stereoscope Greening DB37/T 5084-2016," "Tianjin Roof Greening Technical Regulation DB29-118-2005," and "Qingdao Roof Greening Code." Southern provinces such as Zhejiang, and cities like Shanghai, Shenzhen, Chongqing, Ningbo and Kunming have published the following regulations and measures: "Shanghai Roof Greening Technical Specifications [2015] No. 330,"

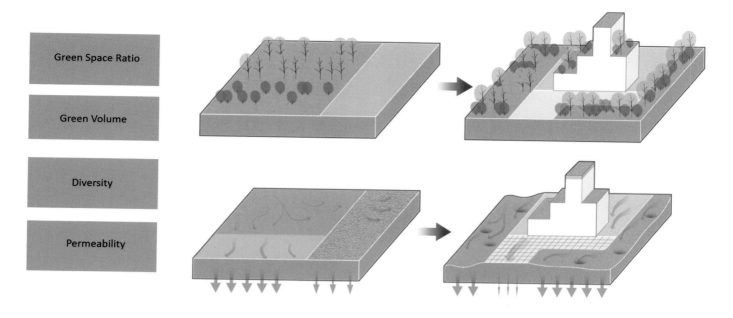

Green Space Ratio

Green Volume

Diversity

Permeability

Figure 1
IUG Group, vertical greening
as a schematic diagram of
ecological urban design.

"(Shanghai) Construction Regulation for Roof Greening DG/TJ08 022 2005," "(Shanghai) Technical Specification for Green Building Planting DG/TJ08-75-2014," "(Shenzhen) Code for the Design of Roof Greening DB440300/T 37-2009," "Shenzhen Three-Dimensional Greening Implementation Measures," "Guideline for Urban Vertical Greening in Zhejiang Province(201902)," "Technical Specifications for Three-Dimensional Greening in Chongqing," "Ningbo Greenbelt Maintenance Technical Regulations," and "Kunming Three-Dimensional Greening Technical Specifications (Trial)."

The above technical regulations are constantly being updated, and a more scientific definition of roof greening types, basic principles, and plant selection is proposed. For new buildings, this includes the expansion and reconstruction of the building roof design technology, standard construction project quality management, maintenance, and irrigation management of the entire work process to establish control indices and standard requirements.

2.2 Vertical Greening Development in Shanghai

From 2011 to 2015, the skyrise greenery sector in Shanghai achieved significant progress in terms of green coverage area, quality, policies, and financial support. In 2007, a clause was added to the "Shanghai Greenery Regulations (Revised 2007)," encouraging the development of skyrise greenery. In 2014, "Guidelines to Promote the Advancement of the Skyrise Greenery Sector in Shanghai" was released by the government, calling for increased efforts for development in the skyrise greenery sector. In 2015, the "Shanghai Greenery Regulations" was published, clarifying the state's stance on the duties, obligations, and rights of developers regarding skyrise greenery. Since 2008, several normative works on the technical aspects have been released, including "Shanghai Skyrise Greenery Technical Guidelines,"

"Handbook on Green Wall Technology," and "Skyrise Greenery Technical Standards." In 2015, Shanghai introduced technical standards, including "Guidelines on Construction and Management of Green Pillars along Highways" and "Guidelines of Skyrise Greenery in New Developments," which have further indicated an increase in political support for the skyrise greenery sector in Shanghai.

By 2016, Shanghai green coverage had reached 143,029 hm², and the green coverage rate was 38.8%. The green space in the built-up area reached 34,104 hm², with 18,957 hm² of parks. The per capita park area increased to 7.8 m² from 7.2 m² in 2012. Shanghai's forestland area covered 111,604 hm², with a forest coverage rate of 15.56%. [4]

The Shanghai Green Management Bureau uses remote sensing to evaluate the location and conditions of the possible rooftops to be greened. Shanghai also encourages the installation of other forms of skyrise greenery, in addition to rooftop greenery. This rapid advancement in the skyrise greenery sector can be attributed to the enforcement of policies and regulations by the relevant authorities. In the Shanghai Vertical Green Project Planning, the target area of vertical greening is 1,200 hm², and the index will be distributed to every district. Considering the increasing economic cost of ground-based urban greening, vertical greening has become an important technical means to improve the ecological environment in high-density urban areas, especially in city centers.

3. Recent Research from IUG on Vertical Greening in Central Areas of Shanghai

Innovative Urban Green (IUG) is a research team for green innovation technology, based at Tongji University. IUG focuses

Figure 2(a) and (b)
Picture of Xuhui Greenland
Being Fun.

on the research and development of experimental projects using three-dimensional garden technology in high-density cities. In ecological urban regeneration design, vertical greening as a renewal method can have important applications in the three cross-scale stages of urban scale–block scale–building monomer. In urban areas, the two scenarios of inventory update and incremental construction are being analyzed in association with the current situation and urban planning requirements. At the same time, urban renewal using three-dimensional greening needs to be examined in relation to the problems faced over the whole life cycle of inspection–assessment–design–construction–maintenance operation. Economic analysis and research on the environmental improvement benefit need to be conducted and used to refine the application of three-dimensional greening. Also, IUG aims to envision scenarios for the future living environment of citizens and the public activity space, propose ideas for daily healthy indoor activity malls and hospitals for citizens' rehabilitation, and digitally combine the management and control of the industrial parks.

3.1 Urban Vertical Planning in Xuhui District

Xuhui District is located in the southwest of Shanghai's central urban area, with an area of 54.93 km². The planned construction land covers about 51.13 km², and the total planned construction area is about 71.48 million m². To illustrate the feasibility of applying vertical greening in urban regeneration, in recent years, research in Xuhui District in Shanghai has focused on the safety and appearance of vertical greening to create novel and attractive roof greening and vertical greening formats (Figure 1).

After the field investigation and verification, the potential green roof area was estimated at 28.63 hm², accounting for about 30% of the remote sensing data. Table 1 shows the survey results for the green roof resources that can be implemented in different streets. According to the results, many vertical spaces have the potential to implement vertical greening.

Newly built projects are examples of land redevelopment through reconstruction, and they are also a form of urban regeneration. The Greenland Being Fun area in Xuhui District is an example of vertical greening design (Figure 2(a) and (b)). A footpath connecting the roof terrace directly from the first floor has been built, combined with a natural landscape, allowing consumers to get close to and experience nature while shopping and relaxing.

In the planning of urban space, vertical greening indicators of ecological urban design, such as the green space ratio, green volume, and bio-diversity, will be combined with the potential application of vertical greening in urban green spaces to research ecological compensation strategies at the block scale in urban areas. Xuhui District plans to complete 15 hm² of new vertical greening in the short term, giving a total of 22 hm². The different indicators of the green roof area that can be implemented on existing buildings in each street, towns and industrial parks, as well as on recent new construction, reconstruction and expansion of building projects are shown in Table 1, in relation to the potential vertical greening resource, street area and the proportion of key projects.

3.2 Evaluation of the Cost Model in the Renewal Projects

Regardless of whether projects are new or renewal of existing buildings, the cost of vertical greening proposals for urban renewal should be evaluated during the renewal, and the costs and benefits should be balanced in terms of ecological economics. The investment cost tends to vary: normally, the renewal cost for existing buildings is

Table 1
Three-dimensional greening
resources survey and
indicators for various streets
and towns in Xuhui District.

Number	Subdistrict	Established green area (hm²)	Available roof green area (hm²)	Planned roof green in short term (hm²)	Planned roof green by 2035 (hm²)
1	Caohejing Street	1.36	0.97	0.50	3.0
2	Fenglinlu Street	0.99	1.60	0.34	2.0
3	Hongmeilu Street	2.42	3.73	0.11	5.0
4	Hunanlu Street	0.40	0.98	0	2.0
5	Huajingzhen Street	0.06	1.45	0.19	5.0
6	Kangjianxincun Street	1.16	0.77	0.02	2.0
7	Longhua Street	0.74	3.37	0.75	2.0
8	Tianpinglu Street	0.72	2.47	0	2.0
9	Tianlin Street	5.71	3.62	0.23	2.0
10	Xietulu Street	2.99	1.01	0.28	2.0
11	Xujiahui Street	3.6	2.27	1.35	3.0
12	Changqiao Street	0.91	1.93	0.15	3.0
13	Lingyunlu Street	0.45	4.46	0.13	2.0
	Total	21.51	28.63	4.05	35

higher than designing vertical greening for new buildings in the architectural design stage. This research is necessary to demonstrate the economic benefits of implementing vertical greening in urban renewal and transformation, especially to balance concerns over costs against environmental benefits in the long term.

During the design process of a building with vertical greening, digitalization with ENVI-met microclimatic data may be a useful reference to optimize the type of vertical and rooftop greening. For example, during the renovation of the roof of a single building, new roof greening was added. Using digital software, different roof planting situations at the same height and the same time were selected for ENVI-met simulation. The roof greening temperature was evaluated for the environmental benefits from four different plant configurations and compared with model calculations. The simulation used plants built-in to the software, as follows: plants xx—lawn; plants sd—high leaf area density, cylindrical deciduous small tree; plants ss—low leaf area density, cylindrical deciduous small tree.

The results showed that different roof planting conditions had obvious differences in the cooling effect of the same site during the hottest time in summer (Figure 3). After planting the green roof, the air temperature distribution became lower. When the roof was bare, 28.25% of the space was above 33.8°C, while for roof plant sd, the area above 33.8°C was only 2.75%. For roof plant xx and plant ss, there was almost no space where the air temperature exceeded 33.8°C. This indicates that the air temperature in the planted plot was generally lower than when the roof was not planted.

The economic assessment of the whole life cycle is important for the implementation of vertical greening. It can show that not renovating existing buildings using vertical greening will lead to greater costs, reduced ecological efficiency, restricted plant growth,

and difficulties in implementing intelligent control drip irrigation measures. Life-cycle cost analysis (LCCA) is a method to evaluate the total cost of buildings and equipment. This data serves as the cost–benefit assessment standard for vertical greenery projects. An assessment formula of vertical greenery can be based on the core data shown above, and refers to the assessment model of green architecture, classified by construction cost:

$$C_{Total} = C_{Design} + C_{Construction} + C_{Finance}$$

Usually, vertical greenery design begins after the architectural design has been completed, and thus it cannot be integrated with the design of the roof and building façades.[5] There is limited capacity for changes to basic architecture elements. The resulting greenery design is often a monotype, and plantings are limited by structural constraints. By comparison, if the design of vertical greenery is undertaken during the base architectural design stage of an urban complex project, the integrated design of vertical greenery is synchronized with the project planning and can be coordinated with the design of the building structure and mechanical and electrical equipment. Green forms can be designed flexibly and reasonably, combined with the roof, façade, and other basic building components. Plant species, growth media, irrigation, waterproofing, and drainage systems can be designed and selected according to different greening requirements. If the greenery designer is involved in the architectural design stage as early as possible, the vertical greenery will be better able to respond to the architectural form of the city complex. The vertical greenery structural system will be integrated with the building so that an appropriate form of greenery, synchronized with the construction of the building,

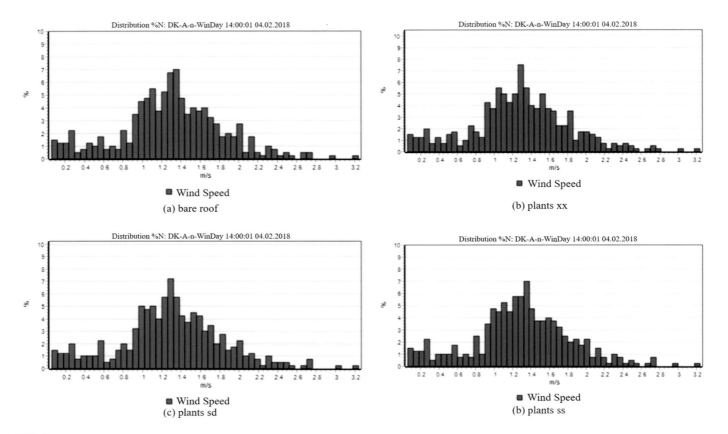

Distribution %N: DK-A-n-WinDay 14:00:01 04.02.2018

■ Wind Speed

(a) bare roof

Distribution %N: DK-A-n-WinDay 14:00:01 04.02.2018

■ Wind Speed

(b) plants xx

Distribution %N: DK-A-n-WinDay 14:00:01 04.02.2018

■ Wind Speed

(c) plants sd

Distribution %N: DK-A-n-WinDay 14:00:01 04.02.2018

■ Wind Speed

(b) plants ss

Figure 3
IUG Group, distribution of outdoor air temperature at H=1.8 m under different roof conditions in Plot A at 14:00 in summer.

Figure 4
IUG Group, the roof of Shanghai Hongqiao Nanfeng City Fengshang Small Farm.

can be considered in cost control exercises. This would reduce the investment cost of urban greenery.

3.3 Economic Benefit of Improvement of the Thermal Environment

Vertical greening also has clear effects in improving the thermal environment. From the perspective of sustainable development, the benefits of roof greening can be divided into three categories, namely ecological, economic, and social benefits. The Joy Garden[6] can be used as an example of the evaluation of the energy-saving benefit of one single building. The IUG team of Tongji University applied various technologies to an experimental roof garden renovation, including discrete measurement of the building structure, waterproof testing and transforming the roof, parametric design of the wooden structure, roof rainwater collection, an automatic irrigation system, and lightweight roof structure matrix within 150 m² of the campus.

This study used a quantitative one-dimensional heat flux calculation method to estimate the energy-saving efficiency and used simulations generated by eQUEST software for comparison and evaluation.[7] The Joy Garden reduced electricity consumption by 1,249 kWh every year on average. Taking 40 years as the whole life cycle, the major ecological benefit is the saving of electricity, accounting for 73.4%; the second largest benefit is the reduction in air pollution, which is 12.2% of the total benefit. For the rest of the benefits, water interception benefit takes 9.5%, and the benefit of carbon reduction appears the least, with a percentage of 9.2%.[8]

3.4 Ecosystem Services from Urban Roof Farms

As an increasingly popular transformation method, rooftop farms can provide multiple ecosystem services, which may be of great help in solving the conflict between urban expansion and agricultural

several aspects affecting the classifying and grading of the roof garden

Classification
• Internal functions------------------------------function
• User groups------------------------------------function

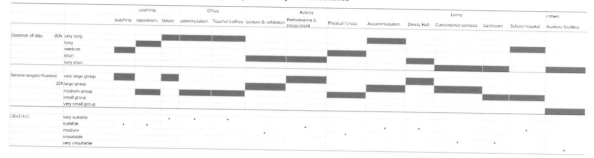

Internal functions	Learning			Office		Activity				Living					Others
	teaching	experiment	library	administration	Teacher's office	Lecture & exhibition	Performance & mega event	Physical fitness	Accommodation	Dining Hall	Convenience services	bathroom	School hospital	Auxiliary facilities
User groups	students teachers staffs others	students teachers staffs others	students teachers staffs others	students teachers staffs others	students teachers staffs others	students teachers staffs others	students teachers staffs others	students teachers staffs others	students teachers staffs others	students teachers staffs others	students teachers staffs others	students teachers staffs others	students teachers staffs others	students teachers staffs others

CLASSIFICATION
roof farming
for leisure & activity
for experiment
more for eco & energy saving

Grading (for people use roof garden)
• Duration of stay --------------------------------using frequency & difficulty of maintenance
• Service range ----------------------------------using frequency & difficulty of maintenance

	Learning			Office		Activity				Living					Others
	teaching	experiment	library	administration	Teacher's office	Lecture & exhibition	Performance & mega event	Physical fitness	Accommodation	Dining Hall	Convenience services	bathroom	School hospital	Auxiliary facilities

Duration of stay — very long / long / medium / short / very short

Service range (influence) — very large group / large group / medium group / small group / very small group

GRADING — very suitable / suitable / medium / unsuitable / very unsuitable

Figure 5
IUG Group, social survey results of use preference in Tongji rooftop gardens.

development. It is difficult for urban agriculture to obtain formal land-use permission in urban areas because of the limited economic benefits. However, in the built environment of high-density cities, new urban agriculture, represented by rooftop farms, is gradually emerging (Figure 4).

The IUG research examines open-air urban roof farms, that is, the green space on the roof of the building that adds productive functions in the city. The purpose of the research is to evaluate the ecosystem services of urban rooftop farms and propose design and planning strategies for urban rooftop farms. Supply (B1), support (B2), social (B3), and regulation (B4) services were used as the basis for the assessment system. Considering that urban rooftop farms are complex ecosystems with more intense human social characteristics, it is important to optimize the indicator system according to the specific characteristics of the urban rooftop farm. In this way, the ecosystem services from urban rooftop farms can be assessed, and the indicators can also be used as the design basis and planning strategy for urban rooftop farms.

According to the model studied in the research, the most important types of ecosystem services were social services, followed by supply services. Owing to the special nature of the rooftop farm planting cycle, the importance of the regulation service is relatively low, and the support service is the least important.

In the future, we can verify the effectiveness of the evaluation system through the analysis of real urban rooftop farms. The detailed requirements of each evaluation indicator will be developed using both qualitative and quantitative methods so that the evaluation model can be applied to the planning strategy of urban rooftop farms. Meanwhile, more attention should be paid to the functionality of social services and supply services in the planning strategy of

urban rooftop farms by increasing the crop planting area, establishing landscape recreation facilities, increasing landscape richness, and improving the frequency of things like science education activities, which can optimize the ecosystem services of urban rooftop farms.

3.5 Social Services of Vertical Greening

Vertical greening also has the function of providing social services in high-density urban environments. The social use demand for green open space by urban residents is an issue that needs to be considered and optimized in the growth of high-density cities. In high-density urban areas, the activity space on school campuses is also restricted to a certain extent. A study investigated the campus usage of Tongji University, finding that teachers and students had high expectations for the potential of roof gardens as an activity space. The comprehensive usage time and main activities of roof garden users were scored, and it was concluded that the roof garden was the most suitable for people who studied and worked on campus for long hours.

There were six factors that led campus users to visit roof gardens according to the survey (Figure 5), and the preferences of different groups of people varied. Students preferred to look at the beautiful views from roof gardens, teachers asked for more abundant activity space, other campus staff considered roof gardens were a peaceful, relaxing environment, and campus visitors were more interested in appreciating the beautiful plants. Students preferred to spend time in the roof gardens in the evening and during the afternoon. Most of the surveyed respondents had little planting experience and were not very familiar with roof garden maintenance, but they showed great interest in participating in roof garden design and expansion, maintenance and management, and even in joining gardening-

Figure 6
IUG Group, rooftop gardens
as social educational sites in
Tongji University campus.

Figure 7
IUG Group, Jinhongqiao
International Center.

related communities.

Because the roof garden in Tongji University has been renovated in recent years, its construction methods, plant species, real-time monitoring, and control processes can be used as important cases of vertical greening practice. These, combined with the teaching practice activities of students, have become an important step in the teaching and learning activities (Figure 6).

4. Discussion of Potential in Future Regeneration

Vertical greening can be combined with the construction and optimization of green buildings, ecological city design control guidelines, and the application of ecological restoration and ecological compensation for green urban areas in the construction of high-density cities in the future.

4.1 Renewal of Commercial Complexes

Vertical greening is encouraged for use both before the construction and after. It is considered to be a beneficial ecological measure at any time. In public space frequently used by urban residents, vertical greening can assist with the renewal of commercial complexes, optimize the daily user experience of citizens in the commercial built environment, and enhance the biodiversity of the indoor environment. For example, Jinhongqiao International Center Roof Garden has a mixture of styles and uses (Figure 7). Horticulture classes, themed salons, parent-child activities, and events are held on the rooftop. Over 175 plants, including more than 50 trees, 400 small trees and shrubs are grown, which helps to enrich urban ecological diversity. Significantly, this case demonstrates that the redevelopment of a commercial complex can make a large contribution to ecological

Figure 8
Healing and exercise environments in hospitals.

greening. It also showcases technical requirements for regulating the quantity of vertical greening and guiding urban renewal design indicators.

4.2 Optimization of the Rehabilitation Environment

Hospitals are the public places that require the healthiest healing environment for faster and better recovery of patients. Vertical greening can provide healing environmental services. The healing landscape provided by vertical greening could also be an important trend and has recently been examined in some projects. The extent to which the renovation and expansion of hospitals could provide a more three-dimensional green healing environment for renovation is also a topic to be discussed.

Landscape, which is an important part of the healing environment of hospitals, has become a focus in the renewal and construction of the hospital environment. Vertical greening provides more spatial potential in the hospital environment. By means of roof greening, additional outdoor public activity spaces that are suitable for rehabilitation exercises and soothing moods have been developed. These can provide support for the outdoor activities of hospital patients. In Shanghai, which faces the rapid construction of public health and medical care, the realization of vertical greening in the reconstruction and expansion of hospitals will provide increased ecological value in the current high-density cities (Figure 8).

4.3 Enhanced Ecological Service Capabilities with Smart Support

The combination of vertical greening and digitalization can achieve precise monitoring and evaluation of benefits, thus supporting future digital ecological city design.

For existing parks, vertical greening can improve the ecological service capabilities of greening on an existing basis. The intelligent equipment for appointments in the park can cooperate with smart management and maintenance to provide support for new three-dimensional greening technologies. Based on the existing practical experience of roof garden monitoring projects, smart technology can complete dynamic real-time temperature and humidity monitoring of the roof environment, including smart drip irrigation. The obtained data can be further applied to calculate the ecological efficiency of the roof environment and provide a good role of support and reference for the construction of future intelligent digital operation and maintenance platforms. The application need not only be limited to one single roof (Figure 9, left); it can also be used for dynamic monitoring and evaluation of buildings such as science and technology parks (Figure 9, right).

5. Conclusion

Vertical greening has become an important means of ecological restoration and improvement in high-density cities. To realize this vision, refined technical support from ecological city and architectural design should be combined. In particular, accurate assessment technology of the environmental benefits and service capabilities of the vertical green space provides a basis for decision-making on ecological renewal under different background goals and different functions. Vertical greening has become one of the new trends in the urban renewal in China, promoting the ecological improvement level of the overall urban environment.

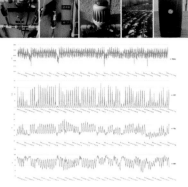

Figure 9
IUG Group, smart monitoring and digital maintenance platform.

Notes

1. Yang He et al., "Thermal and Energy Performance Assessment of Extensive Green Roof in Summer: A Case Study of a Lightweight Building in Shanghai," Energy and Buildings 127 (Sep. 2016), 762-773.

2. Dong Nannan and Wu Jing, "Jiyu ENVI-met moni de wuding lühua rehuanjing yingxiang" [Thermal Environment Effects of Green Roofs Based on ENVI-met Simulation], Journal of Chinese Urban Forestry 18, no. 04 (Aug. 2020), 61-66.

3. Ministry of Housing and Urban–Rural Development of the People's Republic of China, Guojia yuanlin chengshi biaozhun [National garden city standards], 2016.

4. Shanghai Green Management Bureau, 2017.

5. Yang He et al., " Long-Term Thermal Performance Evaluation of Green Roof System Based on Two New Indexes: A Case Study in Shanghai Area," Building and Environment 120 (Aug. 2017), 13-28.

6. The Joy Garden, The Joy Garden Rooftop Experimental Garden was completed in 2016 and is a practical result of the research and design of green roof. It is located on the roof of a 3-story building on the south campus of Tongji University.

7. Dong Nannan et al., "Tongji Daxue shiyan wuding huayuan de sheji he shengtai xiaoyi pinggu" [Design and Ecological Benefits Assessment of Experimental Roof Garden in Tongji University], China Building Waterproofing, no. 23 (Dec. 2017), 19-22.

8. Dong Nannan et al., "Jiyu quan shengming zhouqi chengben–xiaoyi moxing de wuding lühua zonghe xiaoyi pinggu: yi Joy Garden weili" [Comprehensive Benefits Evaluation of Green Roofs Based on Cost–Benefit Model of Whole Life Cycle — A Case Study of Joy Garden], Chinese Landscape Architecture 35, no. 12 (Dec. 2019), 52-57.

Figures

Figure 1: IUG Group, vertical greening as a schematic diagram of ecological urban design.
Figure 2(a) and (b): Picture of Xuhui Greenland Being Fun (Source: https://mp.weixin.qq.com/s/ S8hZrWREODrAJO8BBMBYnA).
Figure 3: IUG Group, distribution of outdoor air temperature at H=1.8 m under different roof conditions in Plot A at 14:00 in summer.
Figure 4: IUG Group, the roof of Shanghai Hongqiao Nanfeng City Fengshang Small Farm.
Figure 5: IUG Group, social survey results of use preference in Tongji rooftop gardens (Source: IUG Group).
Figure 6: IUG Group, rooftop gardens as social educational sites in Tongji University campus.
Figure 7: IUG Group, Jinhongqiao International Center (Source: IUG Group).
Figure 8: Healing and exercise environments in hospitals.
Figure 9: IUG Group, smart monitoring and digital maintenance platform.

Table

Table 1: Three-dimensional greening resources survey and indicators for various streets and towns in Xuhui District (Source: IUG Group).

CH House

CH House and the street

Architect Firm: ODDO Architects
Principal Architect: Mai Lan Chi Obtulovicova, Marek Obtulovic, Nguyen Duc Trung
Design Team: Nguyen Manh Cuong, Quoc Doan Dat
Location: Hanoi, Vietnam
Area: 700 square meters
Completion Year: 2019

Hanoi is Vietnam's capital city and has one of the world's highest population densities, intense traffic jams, frequent air pollution, and a lack of public spaces and greenery, but at the same time, the city finds it of significance in terms of preserving Vietnamese culture and old traditions.

The house is designed for a three-generation family who wanted to create a harmonious space in the hectic city to enhance their sense of traditional family life. The site of the house is a typical plot for the long and narrow local tube houses: in the case of CH House, it is 4.2 meters wide and 35 meters long.

Inspiration
CH House was inspired by the traditional old houses of Hanoi, which have spaces full of natural light and ventilation, thanks to their inner courtyards. The design aims to bring a breath of traditional spirit to modern life, and at the same time, to create spaces full of natural light and ensure natural ventilation within the house.

Space Layout
The functionality of the house is separated into two zones: the commercial area on two floors and the family area. The common spaces of the family area, such as the living room, library, dining room, and kitchen, are positioned on different levels with varying ceiling heights to compose a continuous open space to maintain communication among members of the family more easily. This design makes the space truly open and provides an unexpectedly spacious feeling, despite the limited width of the house. Additionally, the two children's bedrooms are situated above the main common area, next to the void spaces, allowing connection with family members downstairs. The quiet area is located at the innermost part of the house and includes the grandparents' bedroom and the parents' master bedroom.

Significant elements of the interior spaces are three void gaps separating the house's volume to deliver natural light into the lower-level spaces of the house.

Traditional Family
Family ties in the traditional Vietnamese family are very strong. Usually, several generations live together under a single roof where many family events take place. In today's world of modern technology, with smartphones and televisions, these ties have been weakened. The design of the space emphasizes the connections among the family members, especially in the context of today's hurried lifestyle in the new, economically growing Vietnam.

Greenery
Nature is an important element in providing a positive effect on people's mental health. However, the rapid development of large cities creates a lack of green spaces for people to relax. That is why planting trees and plants inside the house is necessary and helps to create a peaceful living space to release stress. CH House is not only a home for dwellers but also a place attempting to create the linkage between human and nature that is very often missing in Hanoi because of its many environmental problems and limited green spaces.

Façade Protection
The façade of the house is designed as a double layer, with an outer layer made from perforated cement blocks and a steel-framed glass inner layer. The double-layered façade, combined with a green layer, provides cover from the sun and dust and allows natural ventilation throughout the entire length of the house. The façade is also designed with a large window, providing even more light to fill the house when needed and giving the façade a more attractive look.

The intention was to design for spatial harmony, taking into consideration family traditions, the local climate, and contemporary lifestyles.

View from the kitchen zone towards the grandparents' room

Diversity of floor and ceiling levels minimise perception of the narrow house layout and enhance natural ventilation.

Double façade made by concrete blocks outside and glazing inside provides shade against harsh sun.

Generous size of the main front façade may positively change perception of the narrow house plan.

View through skylight void and children room towards the front façade where library space is located.

Ground floor commercial zone with a pool for air cooling by evaporation

Commercial spaces in a tiny alley with lush greenery

The surrounding intensive development is apparent from the rooftop terrace.

Light and shadow as remarkable features of the house

Library and relax space

Even a narrow tube house may have diversity of spaces.

Multi-generation family living pattern fostering traditional Vietnamese culture

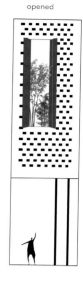

closed   opened

Façade

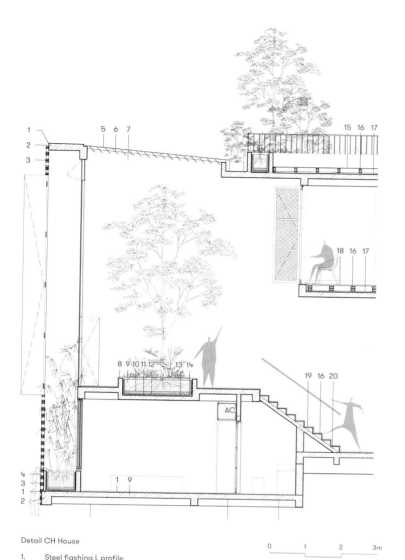

1   5 6 7   15 16 17

18 16 17

8 9 10 11 12   13 14   19 16 20

AC

4
3
1   1   9
2

Detail CH House

1.   Steel flashing L profile
2.   Steel anchor
3.   Gray ventilation bricks
4.   Concrete tree pot
5.   Glass roof
6.   Steel frame
7.   Louvre panels in conrugated iron
      sheet thickness 1mm
8.   Granito tile floor finishing
9.   Soil 400mm
10. Sand layer 50mm
11. Geotextile separation layer
12. Drainage cells 30mm
13. Cement layer 40mm
14. Water proofing insulation 2mm
15. Stone tile floor
16. Bricks
17. Cement layer
18. Bamboo floor mat
19. Granito tile floor
20. Concrete

0   1   2   3m

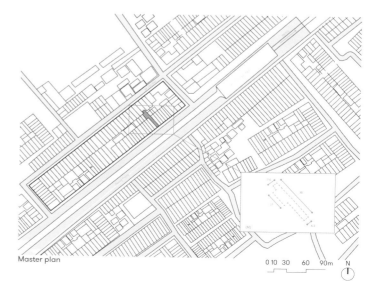

Master plan

0 10 30   60   90m   N

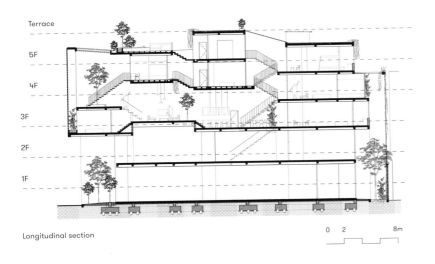

Terrace

5F

4F

3F

2F

1F

Longitudinal section

0   2   8m

The front part of the house with a double layer façade

Airy space layout and visual connections across the house help to achieve better family interaction.

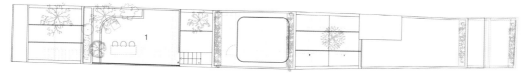

Rooftop floor plan

5F plan

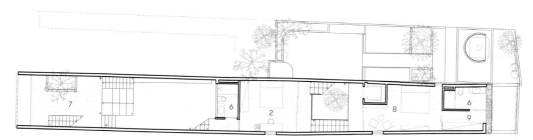

4F plan

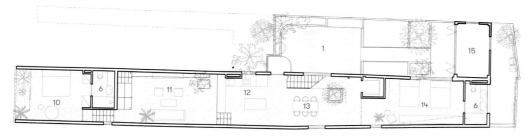

3F plan

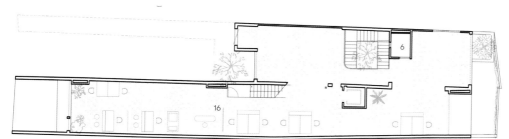

2F plan

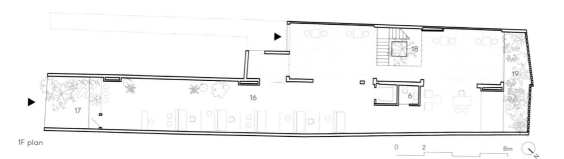

1F plan

1. Terrace
2. Kids' bedroom
3. Washroom
4. Altar room
5. Maid room
6. WC
7. Library
8. Master bedroom
9. Dressing room
10. Guest room
11. Living room
12. Kitchen
13. Dining room
14. Grandparents' room
15. Storage
16. Business
17. Courtard
18. Aquariums
19. Garden

0   2   8m

63

Morning Dew Guesthouse

View from the Chinatown at Choryang

Architect Firm: Architects Group RAUM
Principal Architect: OH Sinwook
Design Team: HA Jeoungun, KIM Daewon, AN Shin, YU Seongcheol, YOON Jeongock, PARK Gyuhyun
Location: Busan, Republic of Korea
Area: 337.58 square meters
Completion Year: 2016
Photography: YOON Joonhwan

This urban guesthouse in Choryang constitutes a new approach to revitalizing tourism and regenerating the old downtown area of Busan, Republic of Korea. With this guesthouse, the landlord will reside in a separate building (having an area of less than 230 square meters), and the guesthouse will offer lodging only for foreign tourists. A notable feature of this project is that the operator is non-Korean and the guests will also be foreigners. I selected this site in Choryang because it is an area where Korean and non-Korean cultures have existed harmoniously. My aim was to create a new type of urban guesthouse, where the foreign customers would be able to appreciate the history and culture of Choryang. After completion, the structure will be open to foreign guests. I hope that this guesthouse will be able to reflect the local characteristics of Choryang.

Choryang is an area of Busan that has played a significant role in the history of that city. In Choryang, most buildings were constructed according to minimum standards, and there is considerable overcrowding in the district. Choryang has long been open to foreign cultures, to which it has adapted as an open port. Thus, Choryang symbolizes an era when the focus was on meeting unplanned daily needs, and it received an influx of non-Koreans over a considerably long period. Various structures have been built in Choryang, and it is possible to encounter both the old and the new while being aware of the connection with the sea. Choryang leaves a strong impression within the port of Busan.

The houses typically found in Choryang are structures that were hastily constructed on the slopes of the district by ordinary residents. Those structures have stood for a considerable time, and today they are regarded as comfortable living spaces. Although the buildings in Choryang are roughly constructed, they follow certain architectural rules, which lends an atmospheric character to this part of the city. Choryang reflects the personality of the people of Busan.

Bird's eye view of Morning Dew Guesthouse

View from the parking lot

Stairs of Morning Dew Guesthouse

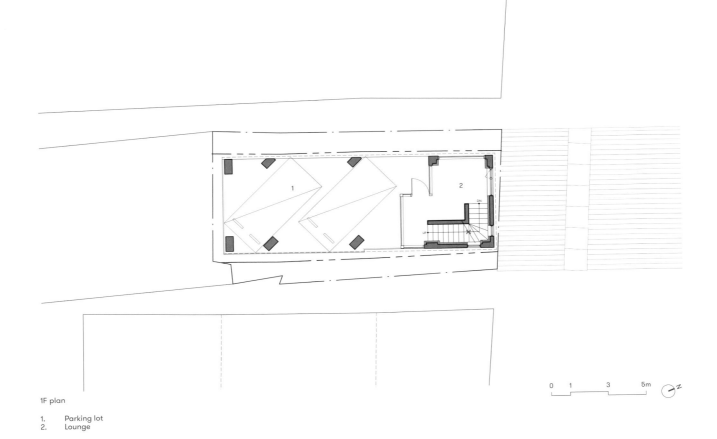

1F plan

1. Parking lot
2. Lounge

0 1   3   5m

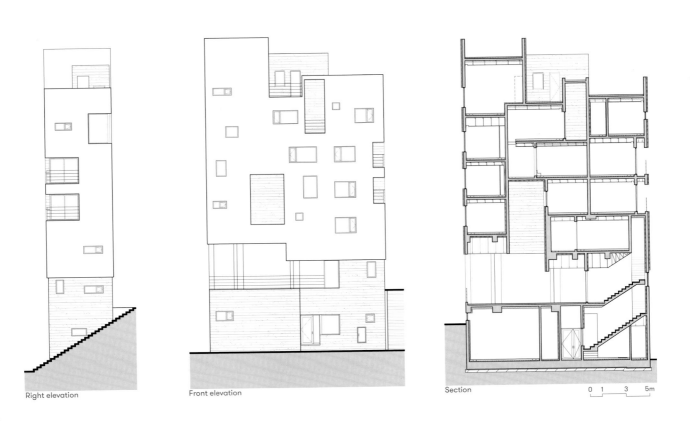

Right elevation

Front elevation

Section

0 1   3   5m

3F plan

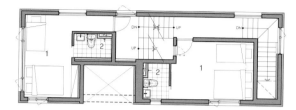

2F plan

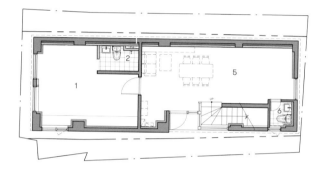

Basement plan

1. Room
2. Bathroom
3. Laundry room
4. Balcony
5. Information

6F plan

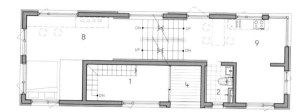

5F plan

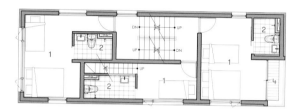

4F plan

6. Toilet
7. Roof
8. Living room
9. Kitchen

Information desk of Morning Dew Guesthouse

Room of Morning Dew Guesthouse

Fai Chi Kei Public Housing

Each of the blocks is centered around a civic square which clusters the apartments to create a sense of neighborhood.

Architect Firm: LBA | Architecture and Planning
Principal Architect: Rui LEÃO, Carlotta BRUNI
Design Team: Nuno ASSIS, Rogério OLIVEIRA,
Arka BISWAS, Jeanie CHU
Location: Macau, China
Area: 699,95.98 square meters
Completion Year: 2015
Photography: António MIL-HOMENS

Extremely high density is the defining characteristic of Macau as a city. The public housing projects, in particular, tend to be of very high density, making it difficult to manage a sequence of urban spaces that move from the public to the private at the ground level. The podium and tower scheme—which shifts the public space to the second- or third-floor level, making it no longer accessible to passers-by—is the typology adopted almost universally in Macau and Hong Kong. Generally speaking, this type of scheme produces a poor quality street interface, because the scale of the tower is not designed with the street block in mind and the podium defines a simplified urban perimeter with minimal sidewalk life.

The Fai Chi Kei social housing project represents an investigation of the potential role of a single building in a wide high-density city context and of the roles of public space within the plot and of the common areas in the overall economy of the project. It is an exercise in the orderly stacking of over 700 apartments and defining a vertical neighborhood.

The Fai Chi Kei is an act of negotiation between the maximum usage of the plot and the livability of the area, proposing a typological alternative to the podium and tower, Hong Kong and Macau's most common housing typology.

The urban footprint of the building presents two large empty patios that gather the housing units and reach the ground, formalizing two open squares where all the lifts, lobbies, and common areas of circulation converge and the shops open their windows. The empty space achieved is a widening of the city walkway: it works both as a seemingly accidental result of urban changes and as a proper entrance square that creates a sense of identity and belonging.

The building also sets out to offer a vertical alternative to the street, because it offers un-purposed common areas at two-floor intervals. These areas are designed with natural light, ventilation and double headroom, creating a breathing element that organizes the community.

The final result is an alternative typology: a porous block with multiple connections on several levels to bring the public space three-dimensionally to the higher floors, thus enabling neighborhood-style living in a vertical setting.

The project explores various systems of stacking to generate social interaction between the different floors.

The towers and the podium are designed as one. The green room and the integrations of public space create a good social environment.

The podium with social equipment is dissimulated inside the tower to allow the volumes to land elegantly on the ground.

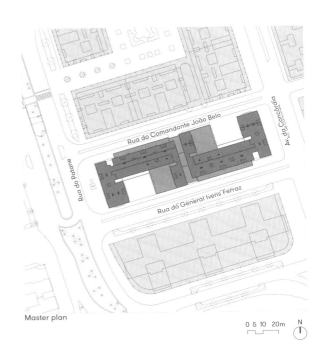

Master plan

0 5 10 20m   N

Sketches

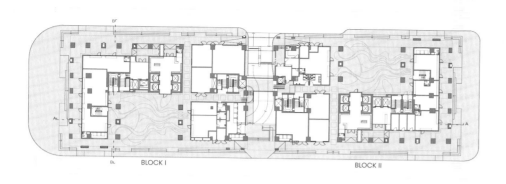

Ground floor plan

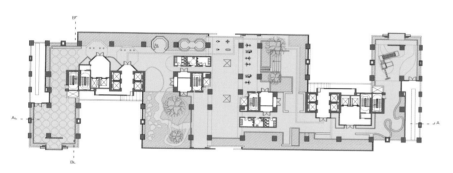

Terrace garden floor plan (3F)

Typical floor plan (4F–21F)

Interior of elevator and main circulation lobby showing the use of double heights and natural light

BLOCK I   BLOCK II

South elevation

BLOCK II

East elevation

0 5 10   20m

Corridor leading to apartments defined by a pattern of grids which allow for natural light to come in, creating a comfortable common space

Double height common patio, where apartment entrances are located, allowing for a space for the habitants to socialize, creating a sense of community

Shanghai Pocket Plaza

Pocket plaza and street

Architect Firm: Atelier Archmixing
Principal Architect: ZHUANG Shen
Design Team: ZHU Jie, LI Lide, DING Xinhui, YIN Jidong, CHEN Hongbang, LU Shuijuan (intern), QIU Xin (intern)
Location: Shanghai, China
Area: 773.86 square meters
Completion Year: 2019
Photography: WU Qingshan, TANG Yu, CHEN Pingnan

The Pocket Plaza is located in a historic downtown district of Shanghai, facing Yongjia Road, on an old street with a pleasant scale and luxurious greenery, surrounded by old residential blocks and small retail establishments.

As part of a regional renewal program, the district government decided to dismantle two rows of shabby residences that represented a fire hazard and to transform the site into a public urban space. For a dense downtown area, we believe an everyday-use open plaza with a sense of privacy and community is what is most needed.

A Small Plaza Enclosed by Galleries
The site is roughly rectangular, 18 meters wide and 40 meters deep, at almost a right angle to the street, enclosed on

three sides by residences, like a pocket. We used open galleries to build a small square, the floor of which is 0.5 meters higher than Yongjia Road, resulting in a sense of territory and an elevated view of the street. A big gentle slope leads people from the sidewalk to the plaza, where the floor is completely covered with red, water-permeable concrete bricks. A hidden fountain serves both as an attraction and a soft management strategy to avoid group dancing, which could destroy the quietness of the place for living. 1.2-meter high shrubs and an iron gate of the same height form a boundary along the pedestrian way. The gate can close at night. At the end of the plaza, two auxiliary rooms are enclosed by zig-zagging, weathered-steel walls for future small businesses.

A Timber–Steel Structure with Dramatic Details
Four galleries form a windmill shape. The ceiling is intentionally lowered to a height of 2.1 to 2.7 meters to build a friendly scale. To enhance the sense of uniqueness and community belonging, we used a timber–steel structure with dramatic details. Each steel pillar is formed of two pieces of flat steel, on top of which stand six layers of wooden beams, stretching out 2.46 meters, fixed with a stainless steel cable at the end, supporting the light metal roof. Wooden benches sit between

concrete volumes holding the steel cables. The contrast between the solid beams' long cantilever and the spatial transparency resulting from the slim pillars and cables further strengthens the sense of structural lightness. We used assembly joints between the beams and pillars, which are rotatable in only one direction to restrain the rotation of the beams and, meanwhile, counteract the force of the wind pushing the roof upwards.

In contrast with the warm tone infusing the surrounding houses with their red bricks, red square floor, timber structure, and rusted steel wall, we painted the steel pillars bright green to create a sense of everyday relaxation.

A Beloved Community Center
A small cafe opened, serving both as an attraction and an invisible "guardian," providing portable seats and preventing inappropriate behaviors. With an open area large enough and plenty of seats, people come here to sunbathe, chat, relax, jog, walk their dogs, and even trim vegetables, while children play in the water fountain. On weekends, Pocket Plaza is decorated to host various activities such as markets, performances, and charity events. This ordinary historic site has transformed into an inclusive meeting point and beloved community center for people of all ages, occupations, and backgrounds.

Original site

Aerial view of Pocket Plaza

The entrance

East gallery and fountain

Public space for residents to chat and rest

Longitudinal sections

0 1 2   5   10m

Lovely fountain

Details of unit structure

West gallery

Cafe behind south gallery

South gallery

Temporary market

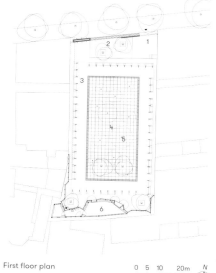

First floor plan

1. Entrance
2. Ramp
3. Gallery
4. Courtyard
5. Fountain
6. Cafe

0 5 10   20m   N

Timber–steel structure

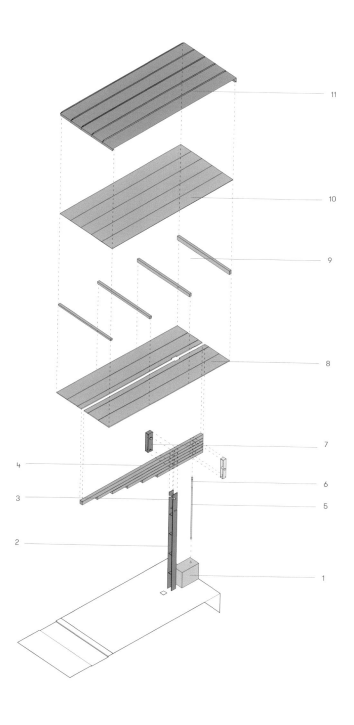

Axonometric analysis of unit structure

1. Concrete block
2. Flat steel columns
3. Latch
4. 70/76 timber tie
5. Steel tie
6. Adjusting rod
7. Steel hoop
8. Roof backboard
9. Timber purlin
10. Roof backing panel
11. Seaming metal roof

Social Labyrinth

Visual connectivity between the inside and outside

Architect Firm: Mobile Offices
Principal Architect: Manisha AGARWAL,
Shantanu POREDI
Design Team: KDPL (Structure) Enova, Cutech,
Epsilon (MEP) Design Ground (Landscape)
Location: Vijayawada, India
Area: 16,441.23 square meters
Completion Year: 2017
Photography: Chintan SHAH

The School of Planning and Architecture campus at Vijayawada is the result of a two-stage national open competition for the master planning of the institute and student housing. For the housing, the intention is to create social spaces that are embedded within their cultural context, along with an environmentally sensitive response to create a built environment that enables learning and living as a community.

The student housing reintroduces traditional principles of architecture and urbanism to contemporary living. The core concept is to focus on the diversity of the individual students and the vastness of the community by creating opportunities for interaction and, thereby, learning. The approach toward the spatial strategy of the student housing program brief is to avoid the hermetic, dormitory-style

organizational structure that tends to foster regimented forms of social control and restrict learning, living, and exchanges. Thus, the housing is envisaged as an active pedestrian ground that simulates a streetscape that is transformed into stilts, verandas, decks, and courtyards within. Three modules with a mix of programmatic and non-programmatic spaces were designed to allow for varied onfigurations around the living courtyards.

The emergent typology has its genesis in traditional tropical responses to climate using passive design strategies. The lower floors are rendered porous by stilts that allow cooler air through the pre-cast block jali (screen) walls along the peripheries. The use of locally available materials such as Tandur stone and a low-rise built form achieves a contextual aesthetic while additionally reducing costs. The project embodies a finer view of sustainability by forming a holistic strategy for responding collectively to various environmental, social, and economic factors.

The building maintains a subtle balance between indoor and outdoor experiences, even though it is a high-density block situated on a constrained site. The design strategies re-envision a new spatial typology for student housing that completely challenges the conventional single- or double-loaded corridor dormitory format commonly adopted for student hostels.

Patterns of light and shadow through the screens

Exterior form made porous with screens which allow for filtering of light and wind

Interconnectivity through terraces and the open spaces

Courtyards create shaded spaces and pedestrian connectivity.

Visually connected decks with interspersed voids that bring in light

At higher level, a greater level of interconnectivity through stairs and the bridges

Relationship of terraces, corridors and stairs around the courtyards

Master plan

GIRLS HOUSING
50m

BOYS HOUSING
45.6m

DINING VISITING BLOCK

0   5   20   35m   N

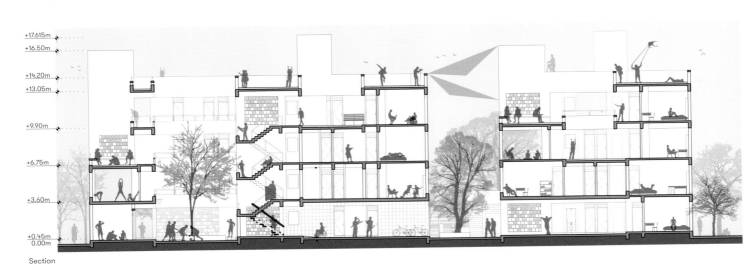

+17.615m
+16.50m

+14.20m
+13.05m

+9.90m

+6.75m

+3.60m

+0.45m
0.00m

Section

84

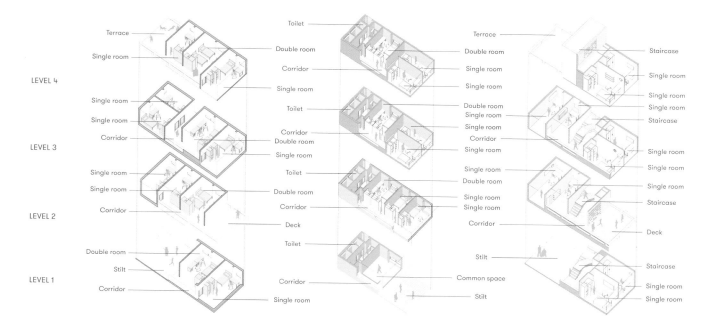

LEVEL 4

LEVEL 3

LEVEL 2

LEVEL 1

Terrace
Single room
Single room
Single room
Corridor
Single room
Single room
Corridor
Double room
Stilt
Corridor
Single room

Toilet
Double room
Corridor
Single room
Toilet
Corridor
Double room
Single room
Toilet
Corridor
Double room
Single room
Toilet
Deck
Corridor
Single room
Common space
Stilt

Terrace
Double room
Single room
Single room
Double room
Single room
Single room
Corridor
Single room
Single room
Double room
Single room
Single room
Corridor
Staircase
Single room
Single room
Single room
Staircase
Single room
Single room
Single room
Staircase
Deck
Staircase
Single room
Single room
Stilt

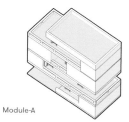

Module-A

8 single rooms + 4 twin sharing rooms +
2 recreational decks + stilt area

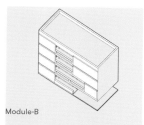

Module-B

4 toilets + 1 common room + 6 single rooms +
3 twin sharing rooms + stilt area

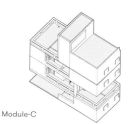

Module-C

Staircase + 10 single rooms +
recreational deck + 1 terrace

Units and modules

Labyrinth of interconnected voids

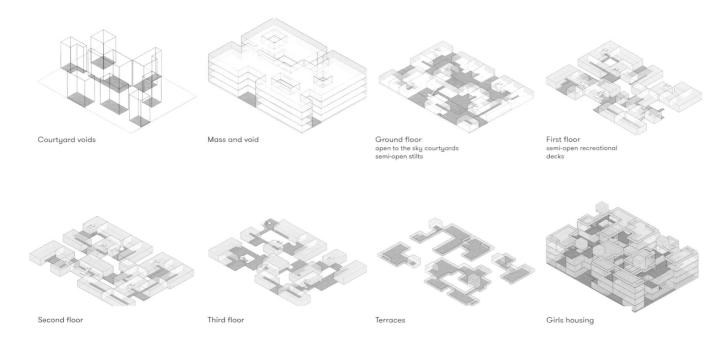

Courtyard voids

Mass and void

Ground floor
open to the sky courtyards
semi-open stilts

First floor
semi-open recreational
decks

Second floor

Third floor

Terraces

Girls housing

Form evolution

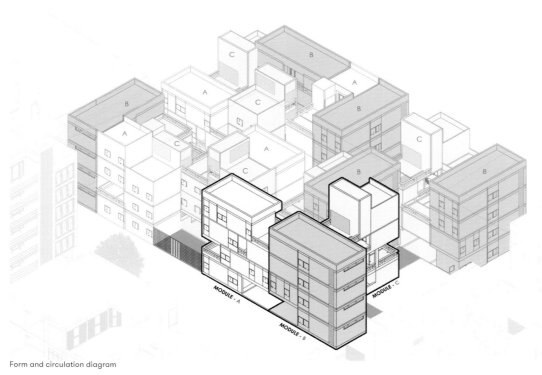

Semi-open stilts

Horizontal and vertical circulation

Terraces and recreational decks

Form and circulation diagram

Three modules with a maximum of programs and non-program have been designed and combined in varied configurations around the living courtyards.

Lianzhou Museum of Photography

The courtyard foyer of the museum opens to the street life.

Architect Firm: O-office Architects
Principal Architect: DONG Jingyu
Design Team: CHEN Xiaolin, LIN Licong, ZHANG Wanyi, DENG Mincong, WANG Yue, HUANG Chengqiang, HE Wenkang, YIN Jianjiang, ZENG Ze, PENG Weisen
Location: Lianzhou, China
Area: 3,400 square meters
Completion Year: 2017
Photography: CHAO Z, Marco CHEN

The idea of building a permanent museum to house professional photographic exhibitions, archives, and educational functions surfaced in 2014, after 10 years of successive international photographic art festivals in Lianzhou, a small town in the north of Guangdong Province, China. The new museum was expected to become the city's social and economic regenerator, a place to unite contemporary arts and local street life.

The project was chosen to be built on the site of a vacant sugar mill in the old city center. It consists of two parts: a three-story original warehouse, transformed into permanent exhibition halls and stockrooms; and enclosing this, a U-shaped new building that houses four exhibition halls of different sizes, a library, a conference hall, and offices. The exhibition space is decentralized, with spaces of various volumes, both in the preserved and newly built parts, which are interconnected by open hallways and staircases that are hung over the ground level to leave spaces open for public events and informal activities. The whole new museum compound is under the shelter of a folded structural roof-façade canopy, a symbol of the local traditional ancestral temple. The sheltered ground-level open spaces are linked with the interconnected semi-open spaces, generating a continuous public spatial flow that connects the front street and the back alley and makes the museum an organic part of the neighborhood.

The museum creates a unique journey experience through the superposition of urban sceneries and contemporary visual arts, from the ground to the rooftop, where an outdoor theater forms the peak of the journey experience. Old pottery tiles, bricks, and wooden windows collected from the demolished buildings have been maintained and reconstructed in the museum. The new museum thus constructs a continuous spatial and material narrative from the city and the old street, creating a new Lianzhou aesthetic in both the archaeological and modernological senses.

The new museum in the old street

Juxtaposition of the new and the old in the neighborhood

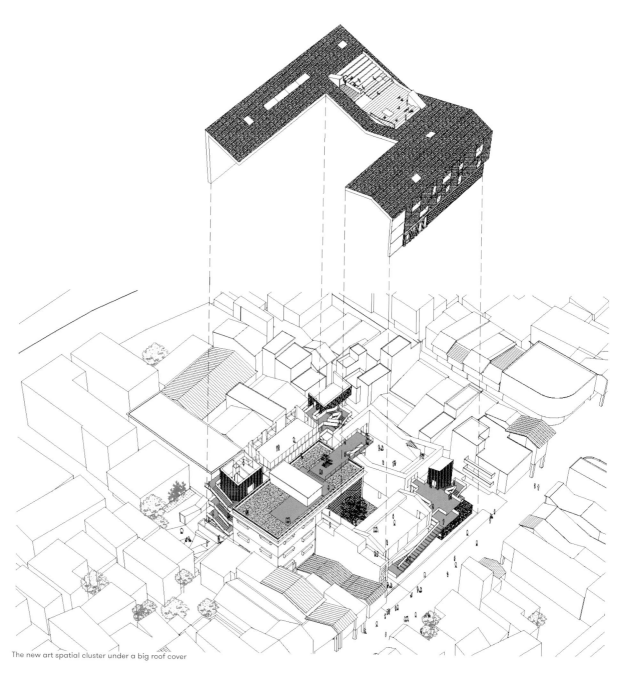

The new art spatial cluster under a big roof cover

Interior of the main exhibition hall under the rooftop theater

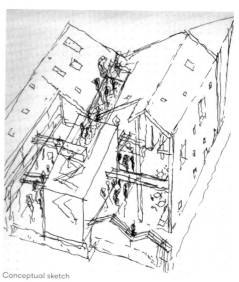

Conceptual sketch

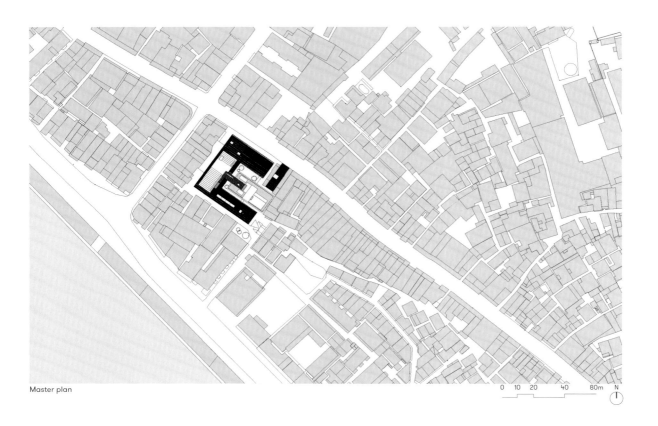

Master plan

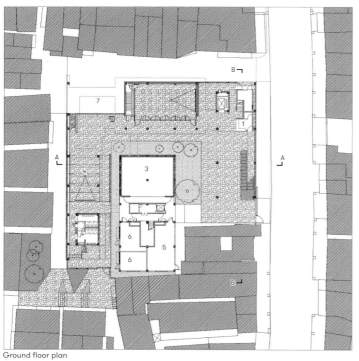

Ground floor plan

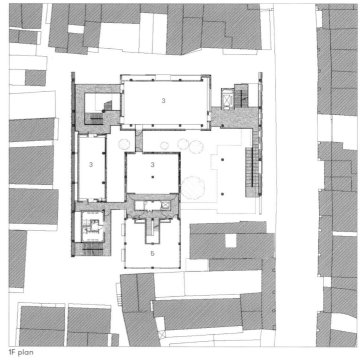

1F plan

1. Reception
2. Multi-function hall
3. Exhibition
4. Outdoor lecture hall
5. Storeroom
6. Equipment room
7. Bicycle area
8. Library
9. Office
10. Rooftop theater

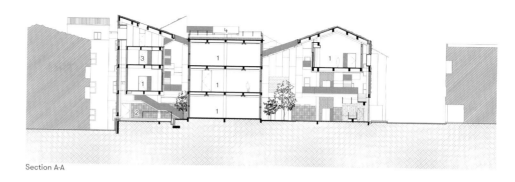

Section A-A

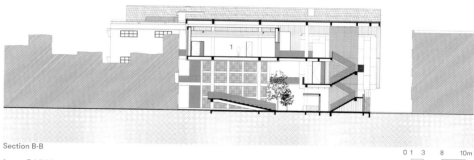

Section B-B

1.   Exhibition
2.   Outdoor lecture hall
3.   Library
4.   Viewing deck

0 1 3   8   10m

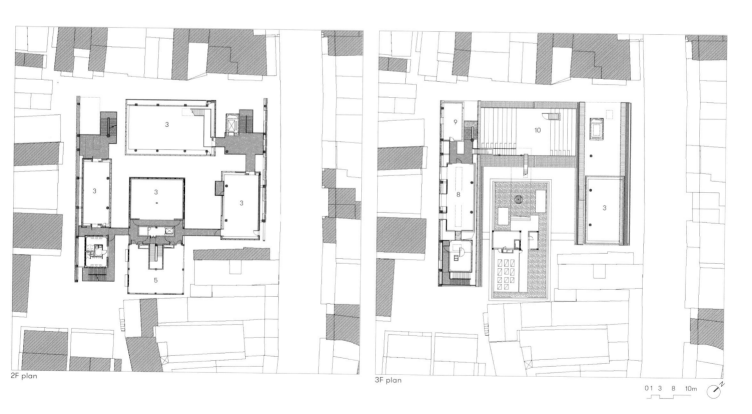

2F plan

3F plan

0 1 3   8   10m

Street façade composed with recycled old wooden windows showcased to the city

New museum rooted in the old city

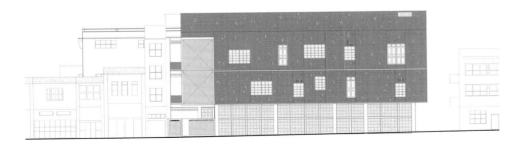

Front (Northeast) elevation

Viewing the old city from the rooftop theater

The ramp rises from ground to the upper exhibition space.

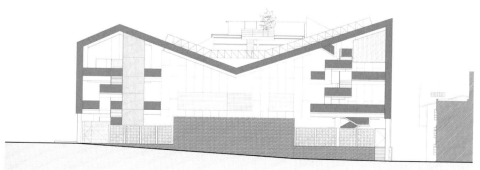
Side (northwest) elevation

The outdoor corridor links the city and the art exhibits.

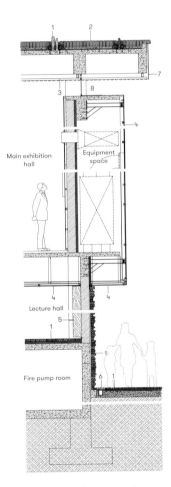

Main exhibition hall

Equipment space

Lecture hall

Fire pump room

Side (Northwest) façade detail

1. Local dark stone
2. Recycled gray shingle
3. PVC corrugated panel
4. Galvanized steel
5. Recycled wooden beam
6. Gravel
7. Black coated steel plate
8. Fixed glazing

The outdoor connection looking over the old city

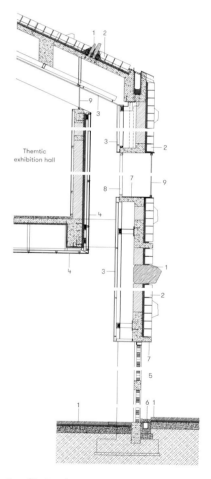

Themtic
exhibition hall

Front (Northeast) façade detail

1. Local dark stone
2. Recycled gray shingle
3. PVC corrugated panel
4. Galvanized steel plate
5. Concrete grass-planting bricks
6. Gravel
7. Black coated steel
8. Recycled wooden windows
9. Fixed glazing

Shenye Tairan Building

Terrace garden

Architect Firm: Zhubo Design
Principal Architect: FENG Guochuan
Design Team: Laura BELEVICA, Aaron ROBIN, LIANG Qibo, ZHANG Zhen, LIU Hailong, LI Jing, MAO Yuanjing, YANG Guang
Location: Shenzhen, China
Area: 168,950 square meters
Completion Year: 2012
Photography: Courtesy of Zhubo Design

100m high and set back from the road by a commercial podium. Instead, the Shenye Tairan Building was conceived as a continuous structure along the edge of the site as close as possible to the road. It extends on the southwestern corner of the site to its maximum allowable height. In that way, it makes full use of its proximity to Binhe Road and Shenzhen Bay. At the opposite northeastern corner, the

building has a lower height and allows easy access for pedestrians to its internal public space.

The result is an urban structure with an inviting vertical courtyard. That courtyard has the tall mass of the building on two sides: it offers a form of protection from both the urban bustle and shade from the hot sun in the south.

Named after its developer, the Shenye Tairan Building stands at a focal point in Chegongmiao, which is an office district for creative industries in Shenzhen, China. The design of the structure is based upon the twin goals of presenting a strong, prominent image within its urban space while remaining approachable and accessible at the human level. To achieve those goals, the designers adopted a fresh approach to creating an office building.

The standard approach to construction on the site would have resulted in two separate office towers,

Bird's eye view

Rooftop garden

Interior view

Bird's eye view

Property line

Lay the footprint of the building close to the edge and set the width of the main volume.

Create a multi-functional courtyard.

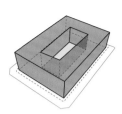

Generate the basic volume according to the total building area.

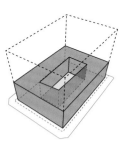

Study the relationship of the volume and the height limit of the building.

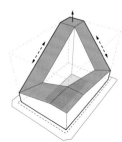

Shape the volume according to the context to maximize the views while responding to the local scale of the neighborhood.

Concept

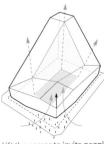

Lift the corner to invite people into the vertical courtyard space — unique room of the city.

Make use of the sloped roof to create a continuous loop of roof garden landscape.

Publicly accessible roof gardens become the 5th façade of the building.

The final shape reminds of the symbolic meaning of "TAIRAN" — strong as the Mountain Tai.

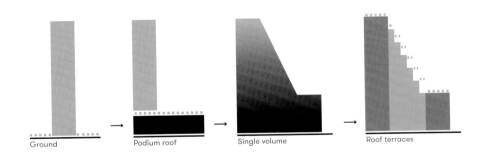

Ground   Podium roof   Single volume   Roof terraces

Green space in an office tower

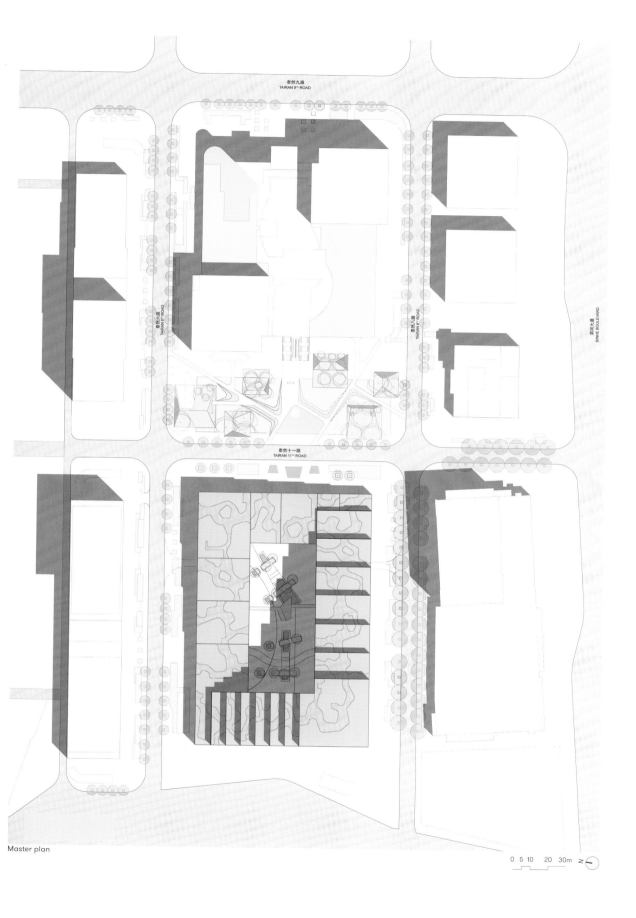

Master plan

泰然九路
TAIRAN 9TH ROAD

泰然六道
TAIRAN 6TH ROAD

泰然八道
TAIRAN 8TH ROAD

滨河大道
BINHE BOULEVARD

泰然十一路
TAIRAN 11TH ROAD

0 5 10   20 30m   N

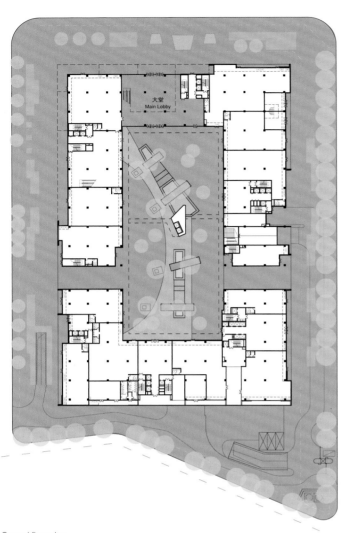

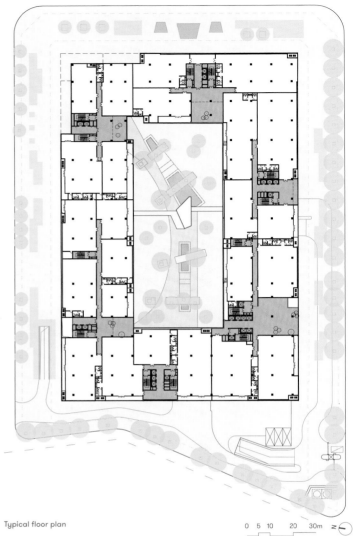

Ground floor plan

Typical floor plan

0 5 10   20   30m   N

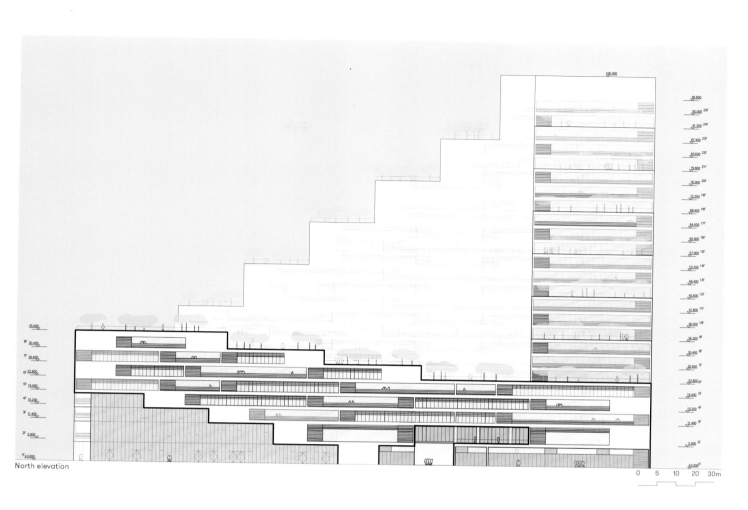

North elevation

0 5 10 20 30m

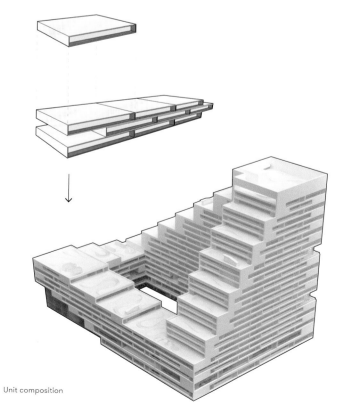

Unit composition

101

Jintan Library

Southeast perspective

Architect Firm: Tongji Architectural
Design (Group) Co., Ltd.
Principal Architect: REN Lizhi, WANG Qiying,
CHEN Xianglei, LIAO Kai
Design Team: RUAN Yonghui, ZHENG Yimin, LIU
Jin, YANG Min, SHEN Xuefeng, TAN Limin, GU
Yong, LIU Yu, WANG Chang, QIAN Daxun
Location: Jintan, China
Area: 15,527.1 square meters
Completion Year: 2017
Photography: ZHANG Yong

The Jintan Library, located on the east side of
the central axis of Jintan's civic square, is
typical of contemporary cultural architectural
practice, featuring an intervention into the
administratively and culturally complex central
district and activating the new urban area,
while remaining open and identifiable in terms
of dimensions of both time and space. It
pays special attention to the local
climate, geography, history, cultural

heritage, and craftsmanship, and departs from
the desire to make the local culture embodied.

On the boundary between the architecture
and the city, a square volume turns into three
interlaced horizontal slices, creating abundant
platforms and covering space. Unlike the
enclosed rest space in a traditional library, the
platforms provide the readers with semi-
outdoor gray space and also block out the
direct sunshine in the interior rooms, acting as
a transition between the square and the
library.

In the interior, the void tree-space grows
from the bottom up, with branches extending
laterally, connecting the exhibition hall below
ground, a training classroom on the ground
floor, and a reading area above ground. The
reader circulation area is combined with a
wandering path through the deformation and
deconstruction of the multi-vocal public center.
The information-sharing space surrounding the
atrium, formed of three cavities at the top, a
terraced area, and a circuitous path, breaks

the monotony of a single center, making the
main circulation space a positive place to visit,
to wander through, or to stay in.

This library integrates its functional
demands in a reasonable manner, avoiding the
blind pursuit of novelty sometimes seen in the
construction of new cities. The final result has
reached a certain degree of quality in both the
exterior appearance and interior space. Apart
from the focus on the continuity of a new
library and the completed government office
building, the design also tries to explore an
abstract, regional, and local expression of the
Wu culture on a simple and rational
rectangular volume.

View of the platform

Entrance view

Detail view from the façade corner

103

View from the entrance square

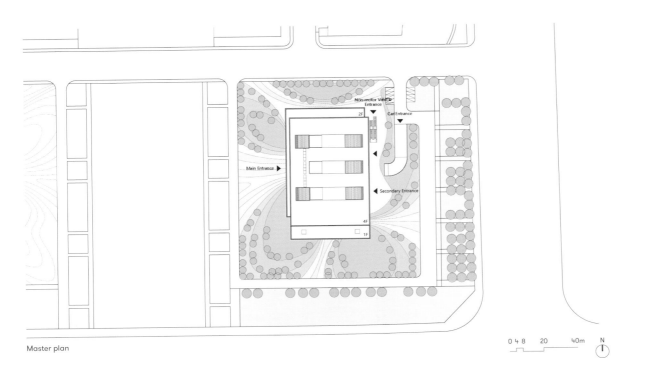

Master plan

0 4 8   20   40m   N

Interior view of the cavity space

Interior view of the central atrium

2F plan

0 2 4 10   20m   N

1. Training room
2. Periodicals reading room
3. Stack room
4. Roof yard

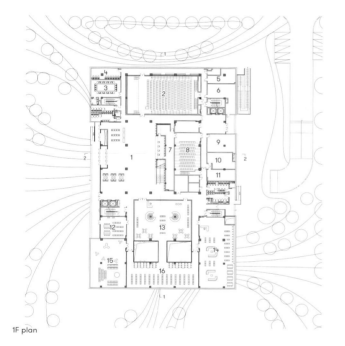

1F plan

1.   Entrance lobby
2.   Conference hall
3.   VIP lounge
4.   Optisch reading room
5.   Storeroom
6.   Fire control room
7.   Logistics room
8.   3D cinema
9.   Editorial processing room
10. Counseling room
11. Information processing room
12. 24-hour public reading room
13. Children's book library
14. Children's periodical reading room
15. Children's playroom
16. Over the underground yard

Basement plan

1. Non-motorized garage
2. Storeroom
3. Yard
4. Exhibition hall

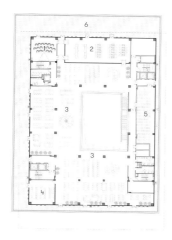

4F plan

1. Equipment room
2. Document clean room
3. Audio and video control room
4. Storeroom
5. Office
6. Roof yard
7. Reference room
8. Export lab
9. Boutique collection room
10. Local literature room
11. Stack room
12. Conference room

3F plan

1. Projection room
2. Multimedia reading room
3. General reading room
4. Study room
5. Stack room
6. Roof yard

Detail view of the bookshelves

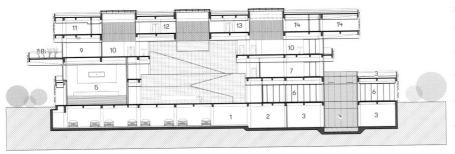

Section 1-1

0 1 3   6   12m

1. Garage
2. Equipment room
3. Exhibition hall
4. Underground yard
5. Conference hall
6. Children's book library
7. Periodicals reading room
8. Roof yard
9. Multimedia reading room
10. General reading room
11. Document clean room
12. Export lab
13. Local literature room
14. Office

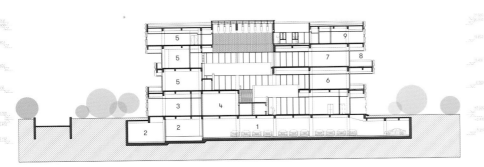

Section 2-2

1. Garage
2. Equipment room
3. Processing room
4. Temporary stack room
5. Stack room
6. Children's training room
7. General reading room
8. Roof yard
9. Boutique collection room

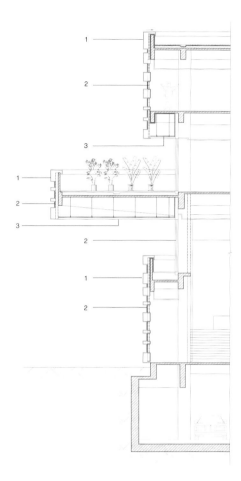

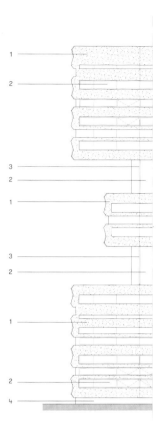

Detail drawing of the wall

1. Off-white stone curtain wall
2. Insulating low-e tempered glass
3. Outdoor light steel keel aluminum ceiling

Detail drawing of the wall elevation

1. Off-white stone curtain wall
2. Insulating low-e tempered glass
3. Dark gray aluminum frame
4. Off-white stone venner

图书在版编目（CIP）数据

亚洲建筑：城市更新 = Architecture Asia：
Urban Regeneration / 亚洲建筑编著 . －－ 上海：同济
大学出版社，2021.6
（亚洲建筑；2）
ISBN 978-7-5608-8649-7

Ⅰ . ①亚… Ⅱ . ①亚… Ⅲ . ①城市建设－亚洲－文集
－英文②建筑设计－作品集－亚洲－现代 Ⅳ .
① TU984.3-53 ② TU206

中国版本图书馆 CIP 数据核字 (2021) 第 115684 号

亚洲建筑：城市更新

Architecture Asia: Urban Regeneration

编　　著：亚洲建筑
出 品 人：华春荣
责任编辑：朱笑黎
责任校对：徐春莲
书籍设计：Telos Books

出版发行：同济大学出版社
地　　址：上海市杨浦区四平路1239号
电　　话：021-65985622
邮政编码：200092
网　　址：http://www.tongjipress.com.cn
经　　销：全国各地新华书店
印　　刷：上海安枫印务有限公司
开　　本：889mm×1194mm　1/16
印　　张：7
字　　数：224 000
版　　次：2021年6月第1版　　2021年6月第1次印刷
书　　号：ISBN 978-7-5608-8649-7
定　　价：68.00元